STREET NAMES
OF
Cleethorpes

Lady Frances Sidney, Countess of Sussex (1531-89), who was the founder of Sidney Sussex College, Cambridge. As the College became the largest landowner in Cleethorpes, it decided on a large proportion of the town's street names.

STREET NAMES
OF
Cleethorpes

Alan Dowling

PHILLIMORE

2010

Published by

PHILLIMORE & CO. LTD

Chichester, West Sussex, England

www.phillimore.co.uk

© Alan Dowling, 2010

ISBN 978-1-86077- 605-2

Printed and bound in Great Britain
Manufacturing managed by Jellyfish Print Solutions Ltd

Contents

Dedicated with love to my children
Michael, Ann, Sarah and Stephen

List of Illustrations

Frontispiece: Lady Frances Sidney, Countess of Sussex

Illustration Acknowledgements

Illustrations are included by courtesy of the North East Lincolnshire Council Library Service (4-5, 7-8, 14, 17-18, 25, 27, 29, 34, 40, 43-4, 46-7, 49-50, 56, 62-3, 67-9, 71-2, 75, 78-80, 84-8, 91, 93, 95-6, 100, 102, 104), the *Grimsby Telegraph* (37, 39, 45, 52, 76-7, 92, 99) and the Cleethorpes Camera Club (12, 28, 41, 57, 101, 103). The frontispiece and illustration 2 have been reproduced by permission of the Master and Fellows of Sidney Sussex College, Cambridge. Maps are reproduced by permission of the Ordnance Survey (64-5), the North East Lincolnshire Archives (23-4) and Rex Russell (1). The remaining illustrations have been provided by private collectors or the author.

Foreword and Acknowledgements

The street names of Cleethorpes are something we take for granted. They are, after all, only the practical means of finding the whereabouts of persons, businesses, buildings or particular locations. So why write a book about them and their origins? The short answer is that a great deal of information about the history and growth of the resort is revealed when we look into the reasons why we have particular street names. Also, many street names conceal stories which are fascinating in their own right – and can add an extra interest to our day-to-day surroundings as we pass a familiar street sign.

The book has arisen out of a long-term interest in local street names. The information contained here has been acquired over many years from a wide variety of sources. These have included books, periodicals, newspapers, directories, council minutes, university records, other archival documents and conversations with local residents. My public lectures on street names have produced information from members of the public, when my own knowledge of a particular street name has proved to be inadequate or non-existent.

Accordingly, I would like to record my gratitude to those many members of the public who have helped in my search for information or who have shared my interest in the subject. I would also like to thank the staff of the Grimsby Central Library's Reference Library, who have all been their usual cheerful and helpful selves during my researches; in particular staff members Jennie Mooney and Simon Balderson, whom I have pestered unmercifully for help. I also wish to thank two archivists for their help: namely, John Wilson of the North East Lincolnshire Archives and Nicholas Rogers, of Sidney Sussex College, Cambridge.

My daughter, Ann Smyth, has been of great help in carrying out computer searches and reading and commenting on the draft texts. Other help has been received from John Houghton, Jean Wales and David Rushworth. I would also like to thank anyone else whose help I may have inadvertently omitted to mention here; and offer my unreserved apologies for the omission. Finally, I am grateful to my wife Dorothy without whose understanding, patience and practical help this book would never have been completed.

Needless to say, I take full responsibility for the content of the book. This statement is particularly significant in a study of street names. The subject follows paths which are strewn with pitfalls, misleading directions and unexpected twists and turns. The origins and meanings of some street names are straightforward (or appear to be so). Others are hidden in the mists of time; whilst some are subject to more than one interpretation or remain an unsolved puzzle. Accordingly, I would welcome gladly any further information on the street names of Cleethorpes.

INTRODUCTION
The Growth of Cleethorpes and its Streets

This book deals with streets within the Cleethorpes town boundary and the few old-established streets which formed the original village of Clee. Clee is now officially in Grimsby but as it was the parent village of Cleethorpes, it makes historical sense to include it here. It should be noted that Clee village now goes by the name Old Clee, helping to distinguish it from New Clee, an area of housing built in Victorian times in the northern part of the Clee parish, between Humber Street and Park Street.

In essence, Cleethorpes street names are a product of the town's historical development. Accordingly, an outline of the growth of Cleethorpes will help to set the subject in context; in particular, by mentioning those events which have led to the physical expansion of the town and the naming of its streets.

The Early History of Cleethorpes

Cleethorpes evolved from three coastal hamlets in the parish of Clee. They were subsidiary settlements, or 'thorpes', of the parent village of Clee. Working from the north, these thorpes were: Oole, which was centred on the present Market Place; Itterby, which was centred on the present Wardall Street and Sea View Street; and Thrunscoe, which lay between the present Segmere Street and the Buck Beck watercourse.

Clee, Itterby and Thrunscoe were listed in 1086 in Domesday Book. Oole was mentioned 30 years later in the Lindsey Survey of 1115-18, with the spelling 'Hol'. By the 16th century, the composite name Cleethorpes was being used for the thorpes of Oole and Itterby. During the 19th century the thorpe of Thrunscoe was absorbed into Cleethorpes.

The early economy of the parish was based on farming, supplemented by fishing, but changes began to take place in the late 1700s as Cleethorpes started to become known as a 'bathing place' for genteel families. Consequently, catering for visitors during the summer season began to supplement local farming and fishing.

In those earlier days the town had few streets and a small number of houses. Accordingly, street names were largely unnecessary. Instead, addresses were frequently indicated by merely using the local terminology for different areas of the town, such as High Thorpe, Middle Thorpe and Low Thorpe (which are explained below in the list of streets). Specific names

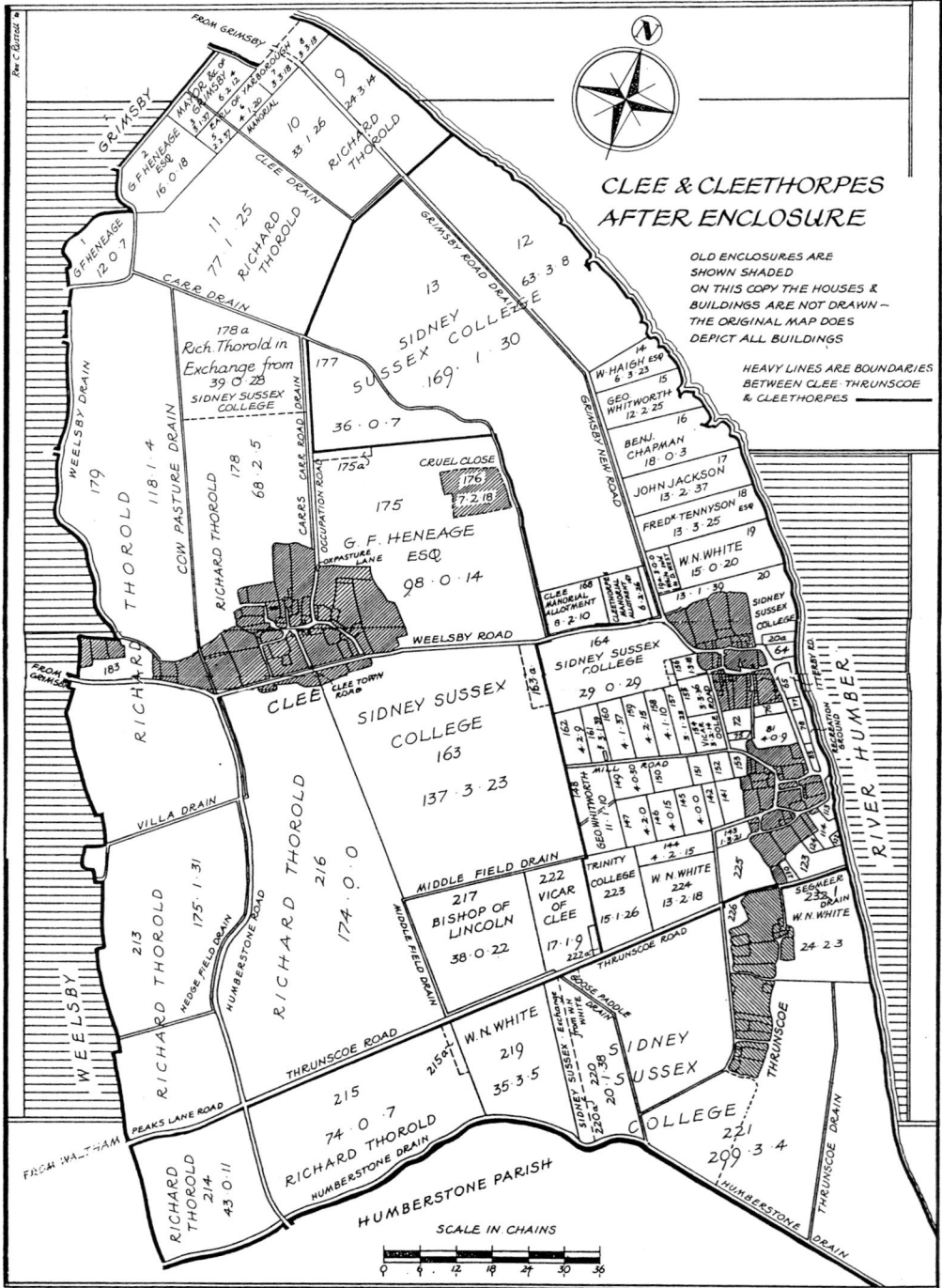

1 *Land ownership in Clee and Cleethorpes in 1846. The shaded areas show the inland village of Clee and (working from north to south) the coastal thorpes of Oole, Itterby and Thrunscoe.*

were used for streets which, for example, indicated an important facility, such as Mill Place, or a main road, such as Itterby Road. Such streets as there were had an irregular 'organic' layout, probably being based on early footpaths, tracks and land boundaries.

Landowners

Landowners would eventually have a major influence on the naming of local streets, particularly Sidney Sussex College of Cambridge University. In 1616, the College purchased an extensive local agricultural estate, the Manor of Itterby. Another major landowner who began to forge a connection with the town in 1787 was the Thorold family, who, by 1798 were living in Clee parish at Weelsby House.

At that time the parish's agricultural land was farmed communally and land belonging to any one owner was scattered in small plots in large open fields. In 1842 the major landowners obtained a private act of parliament, an Enclosure Act, under which an owner's scattered holdings were consolidated into fields which were then enclosed with hedges. Thereafter, landowners had independent control of their land, which they could use for other purposes apart from agriculture, such as building or industry.

The process of enclosure was carried out by an Enclosure Commissioner who also set out (or confirmed) the routes of roads, footpaths and drains and set down their names. His decisions were recorded in 1846 in an Enclosure Award, which will be referred to on occasions in the main text of this book.

At the time of the Enclosure Award, Cleethorpes covered 1,054 acres of land. Sidney Sussex College owned well over half of this land, 603 acres. The Thorold family (who later adopted the surname Grant-Thorold) owned 130 acres and W.N. White owned 115 acres. The remaining 200 or so acres were shared amongst 49 other owners.

The Coming of the Railway

A major event which affected how Cleethorpes, and neighbouring Grimsby, developed and became covered with new streets was the coming of the Manchester, Sheffield and Lincolnshire Railway. Reaching Grimsby during 1848-9, it was extended to Cleethorpes in 1863. This led to a great increase in the number of visitors to the resort, and also an increase in the town's permanent population, who were needed to cater for the visitors. There was also a growing number of Grimsby entrepreneurs who began to find it convenient, and pleasant, to live in Cleethorpes.

All this led to a persistent expansion of the town as more streets were built up and had to be named. Landowners, developers and builders tended to give the new streets names which were associated with themselves in some way. These new thoroughfares were generally given the suffix *street* or *road*, plus an occasional *avenue*.

New Cleethorpes

The town continued to grow as increasing numbers of visitors were attracted by the pier (opened in 1873) and the North and Central promenades (opened in 1885). However, it was the adjacent port of Grimsby which led to the greatest increase in population and street building. By the 1880s Grimsby was running out of building land within easy reach of its docks; such land was needed to house those working in the dock industries. Undeveloped land was available in northern Cleethorpes not far from the docks so, in 1885, Grimsby house building began spilling over the town boundary into Cleethorpes, producing rapid street development in that area, which became, in effect, a dormitory suburb of Grimsby. The suburb became known as New Cleethorpes and covered the area between Park Street and Manchester Street, most of the housing being built on land owned by Sidney Sussex College.

The College had a pronounced influence on the street names used in New Cleethorpes and also in other parts of its Cleethorpes estate. The College policy was that streets on its estate should be named after somebody or something to do with the College – notably College officials, benefactors and distinguished students, but also the names of its estates in other parts of the country. More information on the College's approach to street naming is contained in Appendix C.

Southerly Expansion

Whilst street building was progressing in the northern part of the town, Thrunscoe to the south was in need of protection. The road along the Thrunscoe seafront, Sea Bank Road, was being rapidly eroded and houses along it were in danger of collapsing into the sea. Accordingly, the coast was protected by replacing Sea Bank Road with the Kingsway road and the King's Parade promenade, which were completed in 1906.

The effect of this development was to open up land in Thrunscoe for house building and recreation. An early sign of this was when land lying behind the Kingsway was set out with streets such as Oxford Street and Signhills Avenue. The development of the more southerly parts of Thrunscoe, on Sidney Sussex College land, had to wait until the drainage there was improved.

Municipal recreational development in Thrunscoe began in the 1920s with the creation of the Thrunscoe Recreation Ground; which included the present boating lake and an open-air bathing pool which was replaced by the Leisure Centre in 1983.

Inter-War Expansion

Boundary extensions to the town in 1921 and 1927 roughly doubled its existing size to 2,117 acres. This new land included 583 acres taken in from Humberston; these now contain the golf course and the Country Park. In addition, 392 acres were taken in from Weelsby; these include now the Cleethorpes cemetery and Beacon Hill. But major street-building on this additional land did not take place until after the Second World War. In the meantime, building took place in developing the areas around Sherburn Street and Brereton Avenue; and the building of the town's first council housing estate, centred on St Hugh's Avenue.

After the Second World War

The decades following the Second World War saw the next large-scale expansion of the town – into three main areas. One of them was the large rectangle of ex-Weelsby land bounded by Taylor's Avenue, Humberston Road, Clee Road, Beacon Avenue and Trinity Road. This area includes Beacon Hill and such streets as Sandringham Road, Brian Avenue and Warwick Road.

The second area was in Thrunscoe, between Cromwell Road and the Buck Beck watercourse, including Hardy's Road and the development either side of Chichester Road. The third area was on the ex-Humberston land north of North Sea Lane, including such roads as Seaford Road.

At the time of writing, there are two recent large-scale areas of housing development in the town. They are the on-going Belvoir Road estate between Taylor's Avenue and the Buck Beck and the development south of the Country Park between Park Lane and Rosemary Way.

Changing Street Layouts and Names

The streets of the town demonstrate the changes in layout over the centuries. The largely 'organic' street layout of the original thorpes is shown in the lines of such streets as Cambridge Street. These were then supplemented by the grid-iron layouts of the 19th century and into the 20th century, as seen in New Cleethorpes. The influence of the later 'garden city' concept, leading to more spacious and flexible street layouts, began to be seen in the first Cleethorpes council estate on St Hugh's Avenue.

Since the Second World War there has been a move into more informal, curvilinear street layouts, such as in the Beacon Hill area. This has been accompanied by changes in the naming of streets. Indeed the name *street* has dropped out of general use in new building developments. The names *road*, *avenue*, *place* and *lane* are still used and have been joined by a profusion of *crescents*, *drives*, *closes*, *walks*, *ways*, *mews*, *courts*, *rises*, *paddocks*, *gardens* and other fancies. However, such is the waywardness of fashion that it would not be surprising if the name *street* were to be found making a come-back in the modern retro-style 'village' housing developments.

Sidney Sussex College

As the owner of more than half the land in the town, Sidney Sussex College played a strategic role in the physical expansion of Cleethorpes, and the naming of about sixty of the resort's streets. Accordingly, some information on the College and how it ran its Cleethorpes estate may be helpful to readers.

The governing body of the College was known as the College Meeting. It was headed by the Master, who had overall responsibility for the running of the College. The other members of the College Meeting were the Fellows, who also had individual responsibilities in the operation of the College and its educational function.

The College Meeting had general responsibility for College administration, including the management of its estates and the naming of streets (*see also* Appendix C). The Fellows paid an annual visit to Cleethorpes to inspect the College estate. However, the general administration of the estate was delegated to the Bursar, who was elected from amongst the Fellows. His other functions included responsibility for College finances.

2 *The Fellows of Sidney Sussex College, Cambridge, on their annual visit to Cleethorpes, fitting in a visit to British Titanium Products, c.1959. The Fellows were the governing body of the College and decided on the street names on the College's extensive Cleethorpes estate.*

3 *The Chapel of Sidney Sussex College, where the head of Oliver Cromwell is buried. See the entry for Elliston Street.*

4 *Some of the people who built Cleethorpes streets and houses. Employees of the builders Wilkinson & Houghton in the early 1900s.*

The College eventually withdrew from the town. Its policy had been to develop its land on 99-year building leases but leasehold development became financially unattractive after the Second World War. Accordingly, in the 1950s and 1960s it sold land for freehold building in Thrunscoe on such roads as Howlett Road, Aldrich Road and Hardy's Road. Finally, during 1963-8, it sold all its leasehold property in the town and any land which was still undeveloped. The proceeds were invested in the stock market. This ended the College's long and close association with the town. However, its street-naming policy ensured that it has left a lasting record of its role in helping to create the modern town.

⟶⟹◎⟸⟵

Finally, some words of explanation about the following list of street names may be of assistance to the reader. For instance, the town's larger housing estates, whether council or private, tend to have specific themes in the choice of names. Those thematic street names which are obvious in their derivation are not included in the list. Examples would be streets named after trees, flowers, or birds. Streets which have thematic names which are not so obvious are included in the list.

The question of whether to use an apostrophe in street names is something of a minefield. Sometimes they are used on street signs and elsewhere, and at other times they are not used. Accordingly, if there is any doubt over their use, the apostrophe has been omitted. Abject apologies are herewith submitted for any apparent errors in their use.

The reader will also come across two phrases which may need some explanation. These are used where the local council is said to 'adopt' and/or 'dedicate' a road. In so far as this book is concerned, the former means that the council has taken on responsibility for the physical maintenance of a road. The latter means that a road has been dedicated or 'thrown open' as a public road.

It should also be noted that some entries include information on significant buildings which are now, or have been, in particular streets. Also dealt with are explanations of place-names within the Cleethorpes boundaries, which help to illustrate the history or development of the town. In addition, some names are included which are not in current use but which have an historical or other significance. Other names which are no longer in current use may be found in *The Place-Names of Lincolnshire, Part Five: The Wapentake of Bradley*, by Kenneth Cameron; this is essential further reading for anyone who is interested in the study of local place and street names. A copy may be consulted in the Grimsby Central Library's Reference Library.

Cleethorpes Street Names

ADAMS ROAD

John Adams was Master of Sidney Sussex College, 1730-46. In 1734 there were complaints that he was spending too much on repairs to his College residence, the Master's Lodge. Consequently, the College Fellows voted to limit his unauthorised expenditure to 15s. per half-year.

During his term of office, Bonnie Prince Charlie invaded England in 1745 and got as far as Derby. Accordingly, the College agreed to give £100 to the king, George II, 'for the service of his Majesty on occasion of the present unnatural rebellion'. Adams Road was adopted by the local authority in 1935.

ALBERT ROAD

Many streets are named after royalty. This street was probably named after Prince Albert (1819-61), consort of Queen Victoria. Fish merchant, Walker Moody, lived there in 1871. He was the father of Alderman Sir George Moody who worked hard for the advancement of Cleethorpes.

ALBION TERRACE *see* SEA BANK ROAD

ALDRICH ROAD

Dr Francis Aldrich was Master of Sidney Sussex College, 1608-9. He was one of the original Fellows of the College and died aged 31 after only two years as Master. In 1960 the College decided to set out three new roads: Aldrich, Minshull and Pearson Roads. Several freehold plots for bungalows and houses were for sale at prices from £2,650.

ALEXANDRA ROAD

Originally called Itterby Road because it ran from Oole to Itterby. In 1879, the Prince and Princess of Wales, Princess Alexandra, visited Grimsby. Presumably the road was named after the Princess. In 1896, the road saw the opening of the Alexandra Hall as a public

5   *A Victorian view of Albert Road.*

6   *Oakleigh, 36 Albert Road, 1926.*

venue for concerts, etc. Four years later it was renamed the Empire Theatre and is now an amusement arcade.

The road witnessed tragedy in 1916 when the Baptist chapel was hit by a Zeppelin bomb and 31 soldiers of the Manchester Regiment who were temporarily billeted there were killed. Many others were injured. The dead are commemorated on a monument in the Cleethorpes cemetery. The road was widened in 1939. The public library along the road was opened in 1984 to replace the library on Isaac's Hill. *See also* the entry for Itterby Road.

ALVINGHAM AVENUE

One of the cluster of roads south of Chichester Road with a Lincolnshire theme. Alvingham is the distinctive village north east of Louth which has two adjacent churches and a water mill.

7 Alexandra Road in Edwardian times. The large detached house just left of centre is Yarra House which stood on the corner of Yarra Road. It is now the site of the public library garden.

8 High Cliff, Alexandra Road and the Pier Gardens in the 1930s. Knoll Street is on the extreme left.

9 Pier Gardens and Alexandra Road, c.1950s. The Empire Theatre is just right of centre and the Dolphin Hotel is on the extreme right.

10 The Knoll, built in 1898 on the corner of Alexandra Road and Knoll Street. Seen here in 1994.

AMOS SQUARE

This was part of the closely-packed complex of small houses on what is now the car park in Wardall Street. It was named after Amos, a member of the extensive Appleyard family. In 1850, the square included the following lodging-house keepers: Amos Appleyard, Joseph Appleyard junior, William Appleyard, William Blanchard, John Croft, Solomon Lidgard, Absalom Osbourne and John Priestley. *See also* the entry for Wardall Street.

ANN GROVE

Named after one of the children of Cartledge the builder, along with nearby Brian Avenue, Philip Avenue and Philip Grove.

ANTHONY'S BANK ROAD

This road runs along what was a sea bank called Anthony's Bank. The bank was constructed in the 1790s and probably named after a farm in Humberston called Anthony Farm, which was later named Cottagers Yard Farm. Also, a stretch of salt marsh which lay at the southern end of the bank was called Anthony Water. Nowadays, the road is frequently mis-named *Saint* Anthony's Bank Road but there is no evidence of a saintly connection; it has always been plain Anthony.

ARDEN VILLAGE

One of a cluster of roads off Park Lane in Humberston. Does the name raise visions of the Forest of Arden in Shakespeare's *As You Like It* – or was it merely named after one of the several places in the country bearing this rustic name? There was an Arden Wood at nearby Healing in 1907 when local artist Herbert Rollett featured its trees in a painting.

ARUNDEL PLACE

Most of the roads in this pleasant area of council housing centred on Richmond Road have names based on the theme of castles or stately houses; Arundel Castle is in Sussex.

ASHBY ROAD

Most of the roads in this corner of the town have names with a local or Lincolnshire theme. Ashby is a common place-name in the county, e.g. nearby Ashby-cum-Fenby.

⊷═◉═⊶

BALMORAL ROAD

The names on a large part of the Cleethorpes Council's extensive Beacon Hill estate are on the theme of castles, stately homes, abbeys, etc. Balmoral Castle was built for Queen Victoria after she fell in love with Scotland.

BANCROFT COURT *see* HUMBER STREET

BANCROFT STREET *see* WORKING MEN'S

BARCROFT STREET

William Barcroft was a student at Sidney Sussex College and, more importantly, a College benefactor. In the 1750s he gave the College £800 for the purchase of clerical livings which would be available to members of the College who were clergymen; plus £50 'for glazing and beautifying the College Chapel'. Applications to build 75 houses in the street were put forward in 1893. Work on building the Barcroft Street school began in 1896.

BARK STREET

In 1901, local historian C.E. Watson wrote that Bark Street preserved the identity of the area where the fishermen 'barked' or tanned their sails. Going forward to 1919, there were complaints about the nuisance created by the keeping of swine in the street. Then in 1922, complaints were made about the poor condition of what was at that time a private road; land in the street had recently been sold by Trinity College, Cambridge. The road was eventually 'made up' by the Cleethorpes Council in 1924.

BARKHOUSE LANE

The same derivation as Bark Street, i.e. where fishermen's sails were 'barked' or tanned to preserve them from the elements. It was a private lane in 1921 when the Cleethorpes Council ordered the owners to repair the road. There were no houses in the lane in 1924 and the road was eventually 'made up' by the Council in 1928.

BASSETT ROAD

Joshua Basset (note only one 't') was Master of Sidney Sussex College, 1686-8. He was a Roman Catholic who was forced on to this Protestant College by James II who was committed to reviving Catholicism in the country. Basset was 'a passionate, proud and insolent man wherever he was opposed'. He stole away when James II fled the country in 1688 and left most of his possessions in the Master's Lodge.

BEACON AVENUE

Named after the nearby Beacon Hill but has had several name changes. In 1846 it was known as Millers Road, because of the neighbouring windmill on Mill Road. It was then called Cemetery Road after the cemetery was set out along the road in the 1870s but by 1930 it had its present name.

BEACON HILL

Now absorbed into the extended Cleethorpes cemetery, this ancient barrow (burial mound) is reputed to have been used as the site of a beacon since at least 1377. The land belonged to Sidney Sussex College and the name Beacon Hill was used in a College document of 1734. In 1931, the Cleethorpes Council asked the College to give it the land 'on which beacon fires were lighted as a warning in time of threatened invasion'. This was so it could

Beacon Hill, the ancient barrow (burial mound) in the Cleethorpes Cemetery, seen here in 1989.

be maintained as a place of historic interest. In 1935 the hill was given to the Council and the barrow was excavated under the direction of L.W. Pye, the Council's surveyor, and T. Sheppard of Hull Museums. Bronze-Age remains were found and urns containing Anglo-Saxon cremated bones. At the time of writing the urns are on display in Grimsby's Fishing Heritage Centre. *See also* the entry for the Cemetery.

BEACON HILL ESTATE

This extensive area of building began to be developed in the late 1950s. The roads are treated under their specific names in this list.

BEACONTHORPE

This is the area of Cleethorpes stretching along Grimsby Road from the base of Isaac's Hill and including the Beaconthorpe Methodist Church. It was not one of the ancient thorpes of the town but was named after the shipping beacon which was erected there in 1834 by Hull Trinity House as an aid to navigation. The beacon was on the coast opposite to where Poplar Road now runs. It was demolished in 1863.

BEACONTHORPE ROAD

It is in the Beaconthorpe area of the town and its name was agreed by the Cleethorpes Council in 1929, the same year that the road was adopted.

BECK WALK *and* BECKSIDE CLOSE

Both named after the nearby Buck Beck. *See* the entry for Buck Beck Way for information on the beck.

BEDFORD ROAD

This road, and others in its vicinity, were named after quality London hotels. The other roads are: Berkeley Road, Berners Road, Carlton Close, Cavendish Close, Cumberland Road, Grosvenor Court, Hilton Court, Mayfair Court, Russell Court, Waldorf Road, Westbury Road and Westbury Park.

BEESBY DRIVE

One of the cluster of roads south of Chichester Road with a Lincolnshire theme. Beesby is the site of a deserted medieval village and lies off the ancient Barton Street, the B1431 road.

BELMONT CLOSE

Most of the roads in this corner of the town have names with a local or Lincolnshire theme. The Belmont television and radio transmitting tower is a striking feature on the Lincolnshire skyline, lying a few miles west of the village of Donington-on-Bain.

BELVOIR ROAD

Roads in this part of the resort are named after fox hunts in various parts of the country. This one is named after the Belvoir Hunt, in Rutland.

BENNETT ROAD

Land here was allocated to Benjamin Chapman in the Enclosure Award of 1846. In 1906, Walwyn T. Chapman was given Cleethorpes Council approval to build 36 houses in the road. W. Bennett & Son are listed in the Cleethorpes directory of 1860 as brick and tile makers; whether they had any connection with the road is not clear.

BENTLEY STREET

At the time of the 1846 Enclosure Award, the land over which this street would run was divided between three owners. Starting at the St Peter's Avenue end, they were A.W.T. Grant-Thorold, George Whitworth and Sidney Sussex College. In 1892, plans were drawn up for building that part of the street which was on Grant-Thorold land and which was already named Bentley Street. The following year saw plans approved for building 39 houses. Plans for a further 83 houses

12 *Tree-lined Bentley Street and council houses, in the 1950s.*

13   *Bentley Street council houses, built in the early 1920s, seen here in 1989.*

were approved during 1901–6 and the street was fully built up as far as Wollaston Road by 1906. During the 1920s, the street was extended to Beacon Avenue.

BERKLEY ROAD *see* **BEDFORD ROAD**

BERNERS ROAD *see* **BEDFORD ROAD**

BILLINGHAY COURT

One of the cluster of roads south of Chichester Road with names on a Lincolnshire theme. Billinghay is north of Sleaford.

BISHOPTHORPE ROAD

One of the cluster of roads south of Chichester Road with names on a Lincolnshire theme. Bishopthorpe Farm lies off the road from Humberston to Tetney. There is also a Bishopthorpe south of Whitton in North Lincolnshire.

BLENHEIM PLACE

The names on a large part of the Cleethorpes Council's extensive Beacon Hill estate are on the theme of castles, stately homes, abbeys, etc. Blenheim Palace, in Oxfordshire, was built as a gift from the nation to the Duke of Marlborough after his famous victory at Blenheim in Bavaria in 1704. The Duke of Marlborough was an ancestor of Sir Winston Churchill, who was born at Blenheim Palace.

BLUE MILK TERRACE *see* **NORTH STREET**

BLUNDELL AVENUE

Peter Blundell of Tiverton in Devon was, inadvertently, a major benefactor of Sidney Sussex College. 'From a most humble beginning [he] rose to a great position in the kersey trade [a coarse woollen cloth] and accumulated a vast fortune'. He founded a school at Tiverton and on his death in 1601 left a legacy of £2,000 to Oxford and Cambridge Universities to fund six scholarships for pupils from the Tiverton school. The university authorities allocated most of the legacy to Sidney Sussex College which used it in 1616 to purchase the Manor of Itterby in Clee and Cleethorpes for £1,400. This made the College the largest landowner in Cleethorpes; and proved to be the source of a large income in leasehold ground rents when the land started to be turned over to housing in

14   *Blundell Avenue in New Cleethorpes, 1923.*

the 1880s. Building in Blundell Street began in 1894. Its name was changed to Blundell Avenue in 1908, at the request of the Cleethorpes Council.

BLUNDELL PARK FOOTBALL GROUND

Named after the adjacent Blundell Avenue. In June 1899, Sidney Sussex College accepted an offer from J.H. Alcock of £7,500 for seven acres of land 'for site of proposed hotel and contiguous plot to be used for a football ground'. During the 1899-1900 season, Grimsby Town Football Club moved to the new ground from its Abbey Park ground in Grimsby. The club bought the freehold of Blundell Park in 1927.

BOLINGBROKE ROAD

One of the cluster of roads south of Chichester Road with names on a Lincolnshire theme. The attractive village of Bolingbroke in the Lincolnshire Wolds has the remains of the castle where Henry IV (known as Henry Bolingbroke) was born in 1399.

BOWLING LANE

This busy, narrow, thoroughfare provides a handy direct footway between St Peter's Avenue and Fairview Avenue. It passes the bowling green, which dates from 1906, and from which it takes its name. It is not an ancient lane and was not there in the 1880s. But by 1906 it had been built as far as Glebe Road and the bowling green. It was later extended to Fairview Avenue.

BOY WITH THE LEAKING BOOT *see* KINGSWAY GARDENS

BRADFORD AVENUE

On house no. 2 there is a stone inscribed 'Bradford Terrace 1876', encompassing this short stretch of houses which were the first to be built in what would become the avenue. By

1887 the terrace was referred to as Bradford Street. The street was built up and extended after the Kingsway was opened in 1906. The further extension, between Oxford Street and Hardy's Road, was approved by the Cleethorpes Council in 1908 and was initially called Sheffield Street. The whole street now bears the name Bradford Avenue. The avenue was created a Conservation Area on 13 December 1976.

BRAEMAR ROAD

Presumably named after the Scottish village, home of the famous Highland Games and near the royal residence of Balmoral Castle.

BRAMHALL STREET

John Bramhall (1594-1663) attended school at Pontefract and became a student at Sidney Sussex College in 1609. As Bishop of Derry from 1634 he was a 'stalwart champion of Anglican orthodoxy' and was nicknamed 'Bishop Bramble' by opposing Presbyterians – probably because of his prickly personality? He was a Royalist supporter in the Civil War and exiled himself to the Continent during 1644-60. After the Restoration of the monarchy he was appointed Archbishop of Armagh and Primate of Ireland in 1661. Soon afterwards he became Speaker of the Irish House of Lords. Applications were made to the Cleethorpes Council to build 35 houses and two shops in the street during 1908-9.

BRAMPTON WAY

Most of the roads in this corner of the town have names with a local or Lincolnshire theme. The village of Brampton lies by the River Trent, south of Gainsborough.

15   *Terrace of houses in Bradford Avenue, between Oxford Street and Hardy's Road, built post-1908 and seen here in 2008.*

BRERETON AVENUE

Sir John Brereton was one of the first scholars at Sidney Sussex College when it opened in 1596. He achieved the high-ranking legal office of Sergeant for the Kingdom of Ireland. At his death in 1626 he left the College £2,600. The money was used to buy the estate of Cridling Park in Yorkshire, which brought in an annual rental income of £143. Of this sum, £3 10s. was used to fund an annual commemoration service and feast in his honour. The remainder was used to increase the low incomes of College staff, officers and scholars.

Brooklands Avenue, the enclave off Oxford Street, in 2009. Known in its early days as 'Garden City'.

Building in what was originally called Brereton Street started at the Park Street end in 1894 but by 1906 extended only as far as Sidney Park. In 1908, the College agreed to rename it Brereton 'Avenue', at the request of the Cleethorpes Council. At the same time it asked the Council that the street's future extension should be called Itterby Road, because 'it was as Lords of the Manor of Itterby that the land in the district became their property'. The request was agreed by the Council but then in 1913 the College agreed to a Council request that the name Brereton Avenue should be used for the full length of the road.

BRIAN AVENUE

Named after one of the children of Cartledge the builder, along with nearby Ann Grove, Philip Avenue and Philip Grove.

BRIGHTON STREET *see* NEW BRIGHTON

BROOKLANDS AVENUE

This enclave off Oxford Street was referred to as 'Garden City' in its earlier days. It was squeezed into a triangular piece of land which was left vacant between Rowston Street and Queen's Parade. Its name is reputed to come from a brook which ran into a culvert and thence to the sea. Certainly, the Segmere land drain would have run at one time along the northern edge of the plot – admittedly not quite a babbling brook. One correspondent recalls that in her childhood, the avenue seemed 'almost like a secret magical place to us'. Plans for the first house there, White Lodge, were approved by the Cleethorpes Council in 1911. Building continued piecemeal into the 1930s.

BROUGH COURT

Named after Norman Brough, an 80-year-old tenant of these Royal British Legion flats in Coulbeck Drive; he carried out the official opening ceremony in 1978.

BUCK BECK WAY

Named after the nearby Buck Beck which, in 1298, was recorded as 'bucbek', which comes from the name for a buck or male deer. The beck was the Cleethorpes boundary with Humberston until the Cleethorpes boundary extension of 1921, after which the boundary ran along Humberston Road, North Sea Lane and Anthony's Bank Road.

BURNETT'S AVENUE *see* ELM AVENUE

BURSAR STREET

The Bursar was the Sidney Sussex College officer who was responsible for its finances and the administration of its Cleethorpes estate. Consequently, his importance was worthy of being recognised in a street name. The College started building in this area of the resort in the late 1890s. In 1900, it sold an acre of land for £1,452 for the Bursar Street School, which opened in 1902.

BUTLER PLACE

Dr George Butler was a student at Sidney Sussex College in 1790. He became a Fellow, Tutor and Dean of the College and lectured in mathematics, the classics and ecclesiastical history. He was later appointed headmaster of Harrow school. His son, Dr Montague Butler, also became headmaster of Harrow. Plans for the cul-de-sac were approved by the Cleethorpes Council in 1939.

BUTTERWICK CLOSE

One of the cluster of roads south of Chichester Road with names on a Lincolnshire theme. Butterwick lies between Boston and the sea. Also, the Trentside villages of East and West Butterwick lie a few miles south west of Scunthorpe. East Butterwick was described by Henry Thorold and Jack Yates in 1965 as 'a dull little place … surrounded by root crops and often by fog'.

✦═◗ ◖═✦

CAENBY ROAD

One of the cluster of roads south of Chichester Road with names on a Lincolnshire theme. The hamlet of Caenby is a few miles east of the busy Caenby Corner junction where the road to Gainsborough crosses over the Roman Ermine Street.

CAMBRIDGE STREET

Named after the home city of Sidney Sussex College, which owned a large part of the land where this street runs. The street has had several other names. In the 1846 Enclosure

Award, the section between St Peter's Avenue and Sea View Street was referred to as part of Oole Road. By 1871 it was called Scarborough Street and in 1874 the houses numbered 1-29 were called Cambridge Terrace (there is a stone to this effect above nos 13-15). By 1887, the street bore its present name. According to Methodist historian Dr Frank Baker, the first Primitive Methodist chapel was near to the old pinfold, which was just to the right of where the Town Hall was eventually built. *See also* the entries for Oole Road and Pinfold.

The Odd Fellows Hall was on the street at the junction of the Cuttleby and Yarra Road. It was opened in 1854 by the local lodge of the Manchester Unity of Odd Fellows friendly society. The society had 114 members in 1861 and offered sickness, medical and death benefits. Several of its distinctive gravestones may be seen in local graveyards and the cemetery. The hall was used for various purposes, including meetings of the Cleethorpes local authority prior to the completion of the Town Hall in November 1904. The hall was demolished in 1923 for road improvements.

17 *Cambridge Street, probably in 1944. This terrace of houses was on the right hand side of the street looking towards St. Peter's Avenue. It is now the site of the car park for Town Hall staff which backs on to Oole Road.*

The Town Hall is further along the street. The Cleethorpes Council's first meeting there was held on 15 February 1905. It was originally called the Council House but is now known as the Town Hall. One of the best buildings in the resort, it was designed by the well-known local architect Herbert Scaping (1866-1934). The modern office extension on the seaward side dates from 1987 and has been designed in sympathy with the original Edwardian building.

CAMPDEN CRESCENT

Viscount Campden (1582-1643) became a student at Sidney Sussex College in 1599. He was a nephew of Lord Harington, who was himself a nephew of Lady Frances Sidney Sussex, the founder of the College. Campden was a staunch Royalist and 'raised a regiment of horse' for Charles I during the Civil War but died in the royal quarters at Oxford in 1643. In 1934, the College originally proposed the name Robson Crescent for this new street. The Cleethorpes Council objected, arguing it was too close to Robson Street, which would cause confusion. The College then proposed the present name.

CARLTON CLOSE *see* BEDFORD ROAD

CARR LANE

One of the five roads which meet at the Five-Ways junction. The road is now in Grimsby. However, it was in the ancient parish of Clee and has been included here because of its relevance to the history of the parish. The word 'carr' means boggy lowland, fen or marsh. Carr Lane led to the extensive Clee Carr marsh which lay at its northern end. The Clee karr [sic] was mentioned in documents as early as 1457 and the road was set out in the 1846 Enclosure Award as the 'Carrs Occupation Road' – an occupation road was one that gave access to farm land.

CATTISTOCK ROAD

Roads in this part of the resort are named after fox hunts in various parts of the country. This one is named after the Cattistock Hunt in Dorset.

CAVENDISH CLOSE *see* BEDFORD ROAD

CEMETERY

A pleasant place for a stroll on a sunny day, the cemetery was established in the 1870s. In 1875, the churchyard of the parish church at Old Clee was full and a new cemetery was needed. Sidney Sussex College's offer to sell eight acres of farm land for £1,200 was accepted. In addition, £2,500 was the estimated cost of setting out the cemetery and providing its necessary buildings. As was customary at that time, the cemetery was designed with two chapels. They were designated respectively 'Church of England' and 'Nonconformist'. Only one of them is now used for burial services. The cemetery has been enlarged on several occasions. *See also* the entry for Beacon Hill.

CEMETERY ROAD *see* BEACON AVENUE

CHAPEL LANE *and* CHAPEL ROW

In 1850, both of these locations had lodging houses and their address was simply 'Upper Thorpe', i.e. Itterby. They seem to have been in the vicinity of the first Primitive Methodist chapel which was opened in 1848 on a site to the right of where the Town Hall was later built in Cambridge Street.

CHAPEL YARD

This was located to the left of what is now Steel's restaurant in the Market Place. It was accessed by alleyways from both the Market Place and the High Street. There is still an opening here from the Market Place leading to an open unmade area – which can also be accessed from Cross Street. The first Methodist chapel in the town was opened in Chapel Yard in 1803. There were also houses in the yard and five lodging house keepers lived there. In more recent times, a correspondent remembers it as a dismal courtyard with eight or ten two-up, two-down houses.

18 *Water Tower and Chapman's Pond, in the 1950s.*

CHAPMAN ROAD

In the Enclosure Award of 1846, Benjamin Chapman was awarded 18 acres on the north side of Grimsby Road, in the Beaconthorpe area. The road is named after Walwyn T. Chapman who later opened a brickyard on the land. The water-filled brick pit is known as Chapman's Pond. The road did not begin to be used for housing until the 1920s and was adopted by the Cleethorpes Council in 1928.

CHARLES SQUARE

Charles Square was part of the close-packed complex of houses off Wardall Street – which is now the site of the large car park in Wardall Street. *See also* the entry for Wardall Street.

CHARLES STREET *see* WORKING MEN'S

CHATSWORTH PLACE

The names on a large part of the Cleethorpes Council's extensive Beacon Hill estate are on the theme of castles, stately homes, abbeys, etc. Chatsworth House in Derbyshire, the palatial residence of the Dukes of Devonshire, was reputedly the inspiration for Mr Darcy's house in Jane Austen's novel *Pride and Prejudice*.

CHESTER PLACE

Streets in this small corner assembly are named after prestigious places, such as Chester, which have castles, cathedrals or stately homes.

CHICHESTER ROAD

It must be more than a coincidence that this road was being built and named in the 1960s when Sir Francis Chichester (1901-72) was the English yachting hero of the decade. In 1960 he won the first solo transatlantic yacht race and in 1966-7 he sailed solo round the world – being knighted by the Queen on his return.

CHURCH ROAD *see* ST PETER'S AVENUE

CLAYMORE CLOSE

This street has nothing to do with Scottish swords. Builders Wilkinson & Houghton operated a clay pit and brick works on the site and later built the road and houses there. They decided on the name because the land had produced 'more clay', which when inverted becomes 'Claymore'. *See* the entry for Fairview Avenue for more about the brick pit.

CLEE

The name of the parent village and parish, of which the constituent parts of Cleethorpes were outlying hamlets or thorpes. It was listed in Domesday Book of 1086. The name comes from the Old English word 'claeg' meaning clay or clayey soil, which will resonate

19 *Old Clee Church and buildings, in late-Victorian days.*

with anyone who has a garden in the area. The village is now in Grimsby. This occurred because in 1877 the Local Health District of Clee-with-Weelsby was formed, which included the village of Clee (but not Cleethorpes). So when Clee-with-Weelsby amalgamated with Grimsby in 1889, Clee village and its church became part of Grimsby. This led to cries that Grimsby had stolen Cleethorpes' ancient parish church. The village built-up area is now known as Old Clee. There were 22 families living there in 1563.

CLEE CRESCENT

This street in Old Clee has two outlets on to Clee Road. The crescent-shaped section follows the line of an ancient roadway, whereas the straight section which leads from

20 *Pair of late-Victorian cottages in Old Clee, seen here in 1992.*

Clee Road was set out in 1846 as the 30ft-wide Clee Town Road. It ended at what was called the Town Street, i.e. the 'top' part of the present Clee Crescent. The first Ordnance Survey map to bear the name of Clee Crescent for both sections of the street is that of 1931.

CLEE PARK PLEASURE GROUND *see* PARK STREET

CLEE ROAD

This road between the bottom of Isaac's Hill and Love Lane Corner acquired its present name because it ran between Cleethorpes and Clee village. However, in the Enclosure Award of 1846 it merely formed part of the 30-feet wide Weelsby Road (*see also* the entry for Weelsby Road). In 1934, traffic lights were installed at the Wollaston Road junction. They were described as 'the most modern kind available'; the only other set in Lincolnshire being at Sleaford. Unsurprisingly, two months after they started operating, the police were informed that children were standing on the detector pads to try and make the lights change. This practice was, of course, to become a pastime for children throughout the nation – to little effect.

CLEE TOWN ROAD *see* CLEE CRESCENT

CLEETHORPES

The name is made up of two parts. Firstly, the name of the parent village and parish of Clee; from the Old English word 'claeg' meaning clay or clayey soil. Secondly, the Old English word 'thorp' which signifies a hamlet or village. So 'Cleethorpes' translates as 'the outlying settlements

or hamlets of Clee'. Cleethorpes originally consisted of two thorpes, Oole and Itterby, which were joined in the 19th century by the thorpe of Thrunscoe. In early days the thorpes were referred to under their individual names but documents show that the name 'Clee Thorpes' [*sic*] was possibly being used in 1406. The name 'Clethorpe' [*sic*] was used in 1552 and 'Clethorpes' [*sic*] in 1588. *See also* the entries for Clee, Itterby, Oole and Thrunscoe.

21   *Cleethorpes Town Hall, originally called the Council House, completed in 1904 and seen here in 2008.*

22   *Cleethorpes Cemetery chapels in 1989.*

23   *The central seafront area in 1846 showing land ownership and acreages. The built-up areas of Oole and Itterby are shown to the right and left respectively. Oole Road later became St Peter's Avenue and Itterby Road became Alexandra Road.*

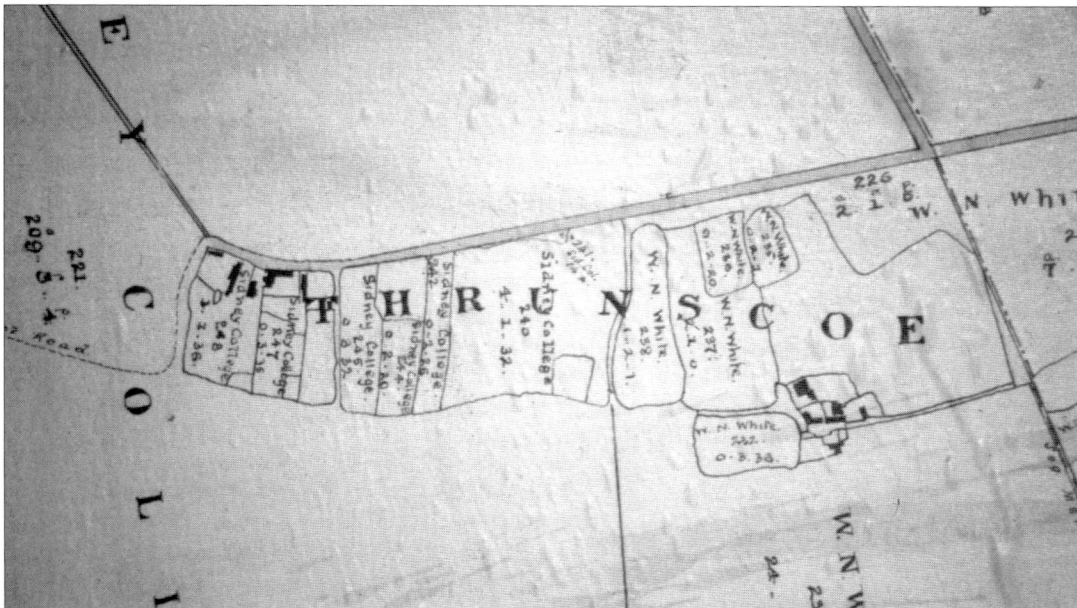

24 *Thrunscoe in 1846 showing land ownership and acreages. The road running across from right to left is now Hardy's Road and the farm buildings to the extreme left are now the site of the Signhills Schools. The road to the right running vertically is now Taylor's Avenue. The buildings of White's Farm are in the lower right-hand quadrant and would be located in the vicinity of the junction of Oxford Street and Bradford Avenue.*

CLERKE STREET

Sir Francis Clerke was one of Sidney Sussex College's benefactors. In 1627 he gave the College land to support new fellowships and scholarships and provide extra accommodation. The College was overcrowded at the time and the income from the land enabled a new wing to be built containing 20 rooms.

CLIXBY CLOSE

One of the cluster of roads south of Chichester Road with names on a Lincolnshire theme. The parish of Clixby is on the edge of the Wolds, two miles north west of Caistor.

CLUMBER PLACE

Most of the roads on this pleasant area of council housing centred on Richmond Road have names based on the theme of castles or stately houses. This one is named after Clumber Park in Nottinghamshire, near Worksop.

COLLEGE STREET (ISAAC'S HILL)

Sidney Sussex College started building on its land either side of Isaac's Hill in the late 1890s – which accounts for the name of this street.

COLLEGE STREET (NEW BRIGHTON) *see* NEW BRIGHTON

COLSON PLACE

Professor John Colson (1680-1760) was appointed Sidney Sussex College's first lecturer in mathematics in 1739. This was a new departure for the College whose main subject of study had always been divinity. The Cleethorpes Council approved the plans for the cul-de-sac in 1939.

COMBE STREET

Francis Combe of Hemel Hempstead was one of Sidney Sussex College's benefactors. He became a student there in 1600 but then moved to Trinity College, Oxford. In 1641, he bequeathed part of his library and 'all that he hath at Abbots Langley and the Lordship there'; to be divided equally between Sidney Sussex College and Trinity College for the education of four of the descendants of his brothers and sisters. In the College's half-year accounts for 1648, the bequest produced £15. Abbots Langley is in Hertfordshire. Building in Combe Street started in 1900.

CONSTITUTIONAL AVENUE

Named after the nearby Constitutional Club, the foundation stone of which was laid in 1911 by the local Conservative Member of Parliament, Sir George Doughty. The club was built for the local Conservative party. For many years 'Constitutional' served as an alternative name for Conservative political ideology; and Conservative Constitutional Clubs were built throughout the country. In 1930, the Cleethorpes Council agreed to plans by builders Taylor & Coulbeck to lay out land in front of the Blundell Park Football Ground and build there 47 houses which would comprise Constitutional Avenue and Imperial Avenue. By 1935, all the houses in the two avenues were occupied. *See also* the entry for Blundell Park Football Ground.

CONYARD ROAD

John Conyard was a brickmaker on Grimsby Road in 1871. The Conyard Brick and Tile Works was located at the end of what is now Conyard Road. It was well established by 1887 but closed prior to the First World War. The site was later purchased by the Cleethorpes Council. The brick pit was filled in and became the site of the Council yard and depot off Poplar Road. Since 2003 it has accommodated a trading estate and also now a medical centre. It is said that during the filling of the brick pit a cart ran backwards into the water with the horse still between the shafts; the horse was drowned. The plans for Conyard Road were approved by the Council in 1911 but the name was not officially ratified until 1928.

CORONATION ROAD

This was originally called White's Lane and led to White's Farm. In 1881, a nuisance was caused in the lane by the 'soakage of foul matter' from adjacent land, which lay higher than the road. One informant remembers it being called Mucky Lane; another recalls the name Dirty Lane. In 1902 the Cleethorpes Council approved the new name of Coronation Road, in view of the pensioners' cottages being built there to mark the accession of

25 *Coronation Cottages in Coronation Road, built to mark the accession of Edward VII in 1901.*

Edward VII. In the 1960s, the road was cut into its present northern and southern halves when Oxford Street was extended to link up with St Peter's Avenue.

COSGROVE STREET

In September 1894, the Cleethorpes Council received plans for a new street, to be called Cosgrove Street. In December of the same year it approved plans for 22 houses to be built there. The street was on land owned by the Grant-Thorold family. The name is taken from Cosgrove Hall in Northamptonshire from where Constance Grant-Thorold was married in December 1904. She was the daughter of A.W.T. Grant-Thorold of Weelsby Hall.

COTTESMORE ROAD

Roads in this part of the resort are named after fox hunts in various parts of the country. This one is named after the Cottesmore Hunt in Rutland.

CRAMPIN ROAD

In 1922, the Cleethorpes Council gave permission for Mr Crampin to build 13 houses on a new road he was constructing off Pelham Road; and also approved the name Crampin Road. In 1935-6, Herbert Crampin, of the well-known trawling family, was living at 385 Grimsby Road (next door but one to the corner of Pelham Road).

CRAVEN ROAD

Dr Joseph Craven was Master of Sidney Sussex College, 1723-8, and died in office. His period in charge passed relatively peacefully but since he had been at the College during the turbulent years of Joshua Basset's Mastership (*see* Bassett Road) he would have appreciated a calm tenure of office. The plans for the street were approved by the council in 1939.

CRIDLING PLACE

Named after Sidney Sussex College's Cridling Park estate in Yorkshire. *See* the entry for Brereton Avenue for more information.

CROMWELL ROAD

Named after Oliver Cromwell (1599-1658), the victor in the Civil War and Lord Protector of the Commonwealth of England. He became a student at Sidney Sussex College in April 1616 just before his 17th birthday. But he left in June 1617, on the death of his father, to take up the life of a country squire. During his time at the College he was more concerned with sport and company than studies. He was 'one of the chief matchmakers and players of football, cudgels, or any other boisterous sport or game'. *See* the entry for Elliston Street for the story regarding Cromwell's missing head.

In 1931, Cromwell Road consisted of a short unmade stretch which ran from King's Road up to a detached house on the northern side built for Mr W.E. Humphreys. The later 1930s saw the road being extended to meet up with Oxford Street (by 1937) and beyond. A spate of building of mainly large semi-detached houses took place between 1933-9, 45 houses being built, chiefly by builder T. Wilkinson.

CROSS STREET

The name of this street is something of a puzzle. It crosses over the High Street and so this appears to be the reason for its name. However, we know that between 1845 and 1887 it was called Cross Street; even though it ran only between the Market Place and the High Street. It did not cross over the High Street until sometime between 1887 and 1906. So the assumption is that it was originally named after someone called Cross. We know that several people of that name lived locally in the Victorian period. It remains a puzzle.

26   *1930s semi-detached houses in Cromwell Road, seen here in 1986.*

27 *Cross Street looking towards the Market Place, 1950s.*

CROW HILL AVENUE

This avenue, built in the early 20th-century off Mill Road, re-used the local name Crow Hill, which had been the old name for Mill Road. Local historian C.E. Watson wrote in 1901 that: 'Crow Hill [Mill Road] the name of the rising ground yet crowned by the old-fashioned windmill is of very doubtful designation'.

CUMBERLAND ROAD *see* BEDFORD ROAD

CUTTLEBY

This footpath between Albert Road and Cambridge Street bears an ancient name, probably of Scandinavian origin. 'Cuttle' could be from a personal name and the 'by' ending is common in this area because of the extensive Scandinavian settlement in the region. It is a Scandinavian suffix signifying a place, homestead or farm. So do we have here evidence of a pre-Norman, Scandinavian settlement? It is perhaps relevant that the name of the hamlet of Kettleby is very similar. Local

28 *Cottages in the Cuttleby, in the 1950s.*

tradition has it that Kettleby, about twenty miles to the west of Cleethorpes, was the residence of Kettel, one of Canute's Danish captains.

With regard to the Cuttleby, variants of the name have been Cuttleby Dale (1734), Cuttleby Close (1739) and Cattleby [sic] Lane (1784). The name has also led to several humorous versions. One was to transform it into 'Cut Belly Lane' – based on the unfounded tale that a gruesome murder took place there. The other is the story of two men moving cattle from one field to another – one man says to the other 'Don't go that way, take them along here – This cut'll be [!] shorter'.

⊷⭻⊙⭺⊶

DAGGETT ROAD

Robert Daggett and Richard Howlett were appointed Fellows of Sidney Sussex College in 1610; and proved to be 'most useful members' of the College. As well as holding tutoring and other posts, the Fellows comprised the governing body of the College.

DAUBNEY STREET

William Heaford Daubney (1816?-95) was the local solicitor and agent for Sidney Sussex College from 1837 until his death in 1895. He played a major role in the development of the College estate, particularly the building of New Cleethorpes in the 1880s and 1890s. He also held the posts of clerk to the local magistrates and registrar of the county court. He was a member of the Grimsby Borough Council, mayor of Grimsby in 1847, 1848, 1851 and 1852 and owned his own estate at nearby Ashby-cum-Fenby.

DAVENPORT DRIVE

Major A.H. Davenport was Bursar of Sidney Sussex College, 1923-55. He was chiefly concerned with College building development in Cambridge and 'the care of the Cleethorpes property'. In 1946, he was 'thrilled and honoured' when the Cleethorpes Council suggested that this new road should be called after him.

DAVIE PLACE

Dr John Davie was Master of Sidney Sussex College, 1811-13. He was in poor health when he was elected Master 'and his death was hastened by his presiding in the Senate House for five hours in bitter weather, in the too conscientious performance of his duties as Vice-Chancellor' of Cambridge University.

DE LACY LANE

The name of this narrow lane off St Peter's Avenue has quite a story behind it, involving the Archbishop of Canterbury. In 1897, the Cleethorpes Council decided that the lane should be called De Lacy Lane. It has been suggested that it was called after Ernest De Lacy Read who was a Clee churchwarden and a manager of the Cleethorpes National Girls' School. In 1886, he and others accused the Vicar of Clee, the Rev. J.P. Benson, and the Bishop of Lincoln, Dr King, with the introduction of ritualistic 'Popish Practices' at the Clee church and the church of St Peter at Gowts in Lincoln. Eventually, the Bishop was summoned to be 'tried' at a court presided over by the Archbishop of Canterbury. The Archbishop's judgement was a compromise between the demands of the High Church adherents and their opponents, who included Read. Read and others then appealed to the Privy Council but their appeal was rejected and the Archbishop's judgement was finally confirmed in 1891.

DOLPHIN STREET *and* DOLPHIN HOTEL

To take the hotel first: it is believed that the original *Dolphin Hotel* was built on the present site *c.*1821 – though its history may go back much further. In 1850 it was advertised as 'This Old-Established Bathing Hotel [which] has for a series of years been in great repute for Families of the first respectability'.

29 The Dolphin Hotel, *probably decorated for Queen Victoria's Diamond Jubilee in 1897.*

30 *Dolphin Gardens, c.1950s.*

It also provided six bathing machines on the foreshore. The present building is the result of a rebuilding of the hotel in 1873-4. It was owned at that time by A.W.T. Grant-Thorold of Weelsby Hall.

The nearby Dolphin Street was named after the hotel and is much younger. In the 1846 Enclosure Award, Richard Thorold's vast estate included a two-acre old inclosure which was three quarters grass and one quarter arable. It was probably in the 1890s that this land was built over with Dolphin Street and the block of buildings along Alexandra Road which include the old Empire Theatre, now an amusement arcade. *See* the entry on Mill Place for the effect of this building work on the windmill.

DRAINS (1846)

Because of the low-lying nature of much of the parish, land drains were an important feature of the local landscape. Detailed descriptions of their courses may be found in the 1846 Enclosure Award, where they are named as follows: Carrs Drain, Carrs Road Drain, Clee Marsh Drain, Cow Pasture Drain, Goose Paddle Drain, Grimsby Road Drain, Hedge Field Drain, Middle Field Drain, Oole Drain, Segmere Drain, Thrunscoe Drain, Villa Drain and Weelsby Road Drain.

DUDLEY PLACE

In 1957, Sidney Sussex College suggested the name Burwash Place for this new cul-de-sac. This was rejected by the Cleethorpes Council, whereupon the present name was proposed. The name is important to the College because Baron De L'Isle and Dudley holds the hereditary position of the Sidney Sussex College 'Visitor'. This came about because of the marriage of Lady Frances Sidney Sussex's brother to the daughter of John Dudley, Duke of Northumberland. Some of the Dudleys also held the title of Earl of Warwick, which may account for a road near Dudley Place being called Warwick Road.

DUGARD ROAD

Richard Dugard was appointed a Fellow of Sidney Sussex College in 1612. He was a leading Tutor at the College during its most flourishing period and was an intimate friend of the poet John Milton. He was also a College benefactor and gave £120 in 1644 to be employed for 'some good permanent use'. He died at Fulletby in Lincolnshire, of which he was rector, in 1653. Plans for the new street, to be called Dugard Road, were approved by the Cleethorpes Council in 1939.

DURHAM ROAD

Streets in this small corner assembly are named after prestigious places, such as Durham, which have castles, cathedrals or stately homes.

EDWARD STREET

This street was largely built-up by 1906. Then the building of the end of the cul-de-sac was completed during 1932-4 when plans were approved for houses to be built by builders Taylor & Coulbeck.

ELLISTON STREET

Dr William Elliston was Master of Sidney Sussex College, 1760-1807. In 1772-3 he was offered the chance to buy the head of the College's most famous pupil, Oliver Cromwell; the seller being 'a comedian'. How this came about was that in 1661, after the Restoration of the monarchy, the bodies of Cromwell and others were dug from their graves, publicly hanged at Tyburn and then beheaded; the heads being then set up on poles on Westminster Hall. Cromwell's embalmed head was blown down during a gale, probably in 1684. Some unknown person, believed to be a soldier, found it and kept it concealed. Elliston declined the offer thinking that the purchase 'might create some prejudice against him'. The head then passed through several hands and was put on show in London in 1775 and 1799. Finally, still in reasonable condition, it was donated to the College in 1960 and buried in a secret location in the College's ante-chapel.

ELM AVENUE *and* ELM ROAD

The name Elm Terrace was included in the 1871 census; Elm Road was named by 1887. In 1894, T.H. Burnett applied for permission to build 10 houses in a new cul-de-sac off the road. The cul-de-sac was originally known as Burnett's Avenue but was given the official name of Elm Avenue by 1923. The bowling green off Elm Road dates from 1906. One tenant of the road must have been the cause of some nuisance to his neighbours. In 1920 he was granted permission by the Cleethorpes Council to carry on the business of rabbit skin dryer. However, in May 1933 he was refused permission to carry on his business of drying skins. But despite this prohibition, it was reported in July that he was still receiving and treating sheep skins.

ELM GARDENS

This short, single-sided, cul-de-sac backs on to Elm Road, from where it must take its name. *See also* the entry for Kew Road.

ESKHAM CLOSE

One of the cluster of roads south of Chichester Road with names on a Lincolnshire theme. Eskham is quickly passed when driving along the coast road between Marshchapel and Grainthorpe.

31 *Demolition of chimney at the brickyard of Wilkinson & Houghton which was where Claymore Close now lies, off Fairview Avenue.*

FAIRVIEW AVENUE

Builders and brickmakers Wilkinson & Houghton built houses in this pleasant avenue in the 1930s. But whoever chose the name must have had his tongue firmly in his cheek because the houses at that time faced the builders' brickpit and brick works. In 1877, a field had been purchased at the top of Mill Road with a view to forming a brickmaking company, but no action seemed to follow. Then in 1900, Wilkinson & Houghton proposed building a brick works there. Controversy followed on the grounds that it would be to the detriment of local residents and the resort generally. Despite this, construction work began early in 1901. The site of the brick pit and works is now covered by Claymore Close.

FAIRWAY COURT

This name was presumably used because of the adjacent golf course.

FAR CLEETHORPES *see* ITTERBY

FILLINGHAM CRESCENT

One of the cluster of roads south of Chichester Road with names on a Lincolnshire theme. Fillingham lies off the road between Lincoln and Kirton-in-Lindsey. In 1970 it was referred to as a 'charming village, with red-roofed stone cottages among trees'. It also has Fillingham Castle, which was built about 1760 in the Gothic style.

FISHER PLACE

Dr Bardsey Fisher was Master of Sidney Sussex College, 1704-23, and was also Rector of Newmarket. His reign as Master was uneventful and the College has portraits of 'this comfortable divine and his handsome wife'. But the use of his name for Fisher Place was far from welcome. The Cleethorpes Council agreed to the name in 1933. However, in 1935, there was a petition from residents asking for it to be changed. The College refused the request, replying that the name was already on all the house deeds and a change would cause confusion.

FISHERMAN'S ROAD

This was the name of the southern part of High Cliff Road, between Humber Street and Brighton Street. In the 1846 Enclosure Award it is described as: 'A Bridle Road of the breadth of Twelve feet called the "Fisherman's Road" commencing at Raven Leys ... and extending in a Southwardly direction ... to the shore of the Humber'. *See also* the entry for Raven Leys.

FITZWILLIAM MEWS

Roads in this part of the resort are named after fox hunts in various parts of the country. This one is named after the Fitzwilliam Hunt near Peterborough.

FIVE-WAYS

Although this road junction is not in Cleethorpes, the name Five-Ways was suggested to the Grimsby Council by the Cleethorpes Council in 1937 when the newly-built Queen Mary Avenue (in Cleethorpes) became the fifth road to meet at the junction.

FOLLY HOLE

This is an old name which is still used by some residents when referring to the Sea Road area. The Folly Hole was a track which was replaced by the Sea Road as part of the railway company's seafront improvements of 1883-5. Two origins of the name have been suggested. One comes from C.E. Watson in his local history book of 1901. He proposed that the name came from the Anglo-Saxon phrase 'ful-ea' or 'Dirty Water Hole, the hollow through which the Oole surface sewer of old precipitated itself into the Humber'.

The other, more attractive, origin can be found in Edward Dobson's guide to Cleethorpes published in 1850 and 1858. Dobson described how popular gatherings, known as 'Fortnight Sundays', took place every other Sunday during the summer in the area of the present Sea Road. The crowds that attended could watch or take part in such activities as boxing, wrestling and football or spend money at stalls which sold foodstuffs and other goods. Alcohol was also on hand and the event became notorious for scenes of drunkenness and rioting. Also, people believed it desecrated the Sabbath and, therefore, described it as a Folly, or the Folly Feast. Thus the area where it was held became known as the Folly Hole. Dobson writes that public opinion finally caused the magistrates to ban the Folly Feasts entirely from 1786.

FOOTPATHS (1846) *see* APPENDIX B

FORE THORPE *see* OOLE

FRANKLAND PLACE

Dr John Frankland was Master of Sidney Sussex College, 1728-30. He was also Dean of Gloucester Cathedral and, later, Dean of Ely. He was elected Master because the Fellows' first choice for the position (the Rev W. Chambre) was quite comfortable in his Yorkshire rectory and turned down the offer. Being Master during this period was not an appealing proposition as the College presented a dismal picture of tottering buildings and financial depression.

Even so, Frankland was not the College's first choice of name for this cul-de-sac. In 1933, the College suggested Chafy Place (named after another late Master of the College); but because of local complaints about the College using unusual street names, it was changed to Frankland Place. *See also* Appendix C for more information on this change of name.

FREDERICK STREET *and* GILES STREET

These two streets, either side of Bursar Street School, were built as part of Sidney Sussex College's development of this area which started in the late 1890s. The published history of the Bursar Street School proposed that Frederick Street was named after the school's architect Frederick Croft. However, the College policy was to give its own names to streets on its land. Accordingly, it is more likely that the street was named after Sir John Frederick who was a College benefactor. He left £1,000 to the College in 1755 for the purchase of church livings. It was put into a College fund for this purpose and in 1765 was used to help purchase the living of Gayton in Northamptonshire. At the time of writing, the source of the name of Giles Street is a mystery.

FREESTON STREET

John Freestone [*sic*] of Altofts, Yorkshire, was a Sidney Sussex College benefactor. He gave £500 for buying an estate to support a Fellow and two Scholars at the College – the Fellow to be a 'Yorkshire man born'. The money was used in 1607 to buy land and houses near Stamford which yielded £25 in annual rents.

FULLER STREET

Dr Thomas Fuller (1608-61) joined Sidney Sussex College in 1629 as a Fellow Commoner. He was renowned for his wit and wisdom as a preacher and writer. As a Royalist during the Civil War he joined the king's court in Oxford in 1643. He took part in the defence of Basing House in Hampshire in 1643-5, during which he headed a sortie against the Parliamentarian trenches. He was known as one of the 'great Cavalier parsons'.

✦⊶◉⊷✦

GARBUTT PLACE

Richard Garbutt was appointed a Fellow of Sidney Sussex College *c.*1612 and became an 'excellent' College officer. His namesake, James Garbutt entered the College in 1616, the same year as Oliver Cromwell. He became a Fellow and a well-known Puritan preacher.

GARNETT STREET

One-eyed John Garnett (1709-82) became a Fellow of Sidney Sussex College in 1730. He left Cambridge in 1751 and went to Ireland as Chaplain to the Lord Lieutenant. He was appointed Bishop of Ferns and later of Clogher. He published a Dissertation on the Book of Job in 1749 which was long-winded 'to a degree which would have taxed all Job's patience and surpasses ours'. He was described as 'a prelate of great humility and a friend to literature and religion. Though he had but one eye, he could discern men of merit.'

THE GATHERUMS

This small pleasant estate, built in 1998 off Beacon Avenue, has revived an old local name. The land where the houses have been built was once known as 'The Gatherums'. Although the name may sound like a life-threatening Victorian ailment, the truth is that in Lincolnshire dialect, Gatherums signifies a passage or narrow road leading from one field to another; it can also mean the entrance to a field. Such a lane is described in the 1846 Enclosure Award and is also shown on the Enclosure Plan. So, over time, the name of the lane has transferred to the land through which it passed. There are other Gatherums in Lincolnshire, including one in nearby Louth. Variant spellings which have been recorded include Gatrum and Gatteram. The derivation of the word is from the Scandinavian gata, meaning street or road, such as is found in many of neighbouring Grimsby's ancient street names, for example Wellowgate, meaning Wellow Street. By the way, please pay no attention to those who suggest The Gatherums name arose because it was 'a gathering place' for local residents – the suggestion is simply is not true and merely a piece of local romantic fantasy.

GAYTON ROAD

One of the cluster of roads south of Chichester Road with names on a Lincolnshire theme. Take your choice from Gayton-le-Marsh or Gayton-le-Wold, both not too far from Louth.

GEORGE STREET *see* WORKING MEN'S

GILBERT SUTCLIFFE COURT

Gilbert Sutcliffe was the Cleethorpes Town Clerk, 1935-74. He was made an honorary freeman of the borough in 1974. He died in 1988.

GLEBE ROAD

This road takes its name from the fact that it was built on glebe land – that is land which is held by the vicar or rector of a parish. In this case the land was awarded to the vicar of Clee in the Enclosure Award of 1846. Part of the same glebe land was used as the site for St Peter's Church.

GORING PLACE

George Goring (1583?-1663) became a student at Sidney Sussex College in 1600. He later served in Flanders and was knighted in 1608. He had a rare gift of wit and drollery, became a favourite courtier of James I and was described as 'Master of the games for fooleries'. Charles I was also fond of him and he became by royal favour the holder of lucrative monopolies. He held the monopolies of tobacco and the licensing of taverns and in 1641 his income was £26,000. He spent his money freely in the King's service, saying: 'I had it all from his Majesty and he hath all again'. He was created Earl of Norwich in 1644. After the Civil War he was condemned to death by Parliament but his sentence was later respited on the casting vote of the Speaker. After being set at liberty he joined Charles II on the Continent. At the Restoration of the monarchy he was appointed Captain of the King's Guard. He was buried in Westminster Abbey. Goring Place was adopted by the Cleethorpes Council in 1931.

GRAINSBY AVENUE

Most of the roads in this corner of the town have names with a local or Lincolnshire theme. Grainsby is an attractive hamlet which hides off the A16 road to Louth.

GRANNIES PASSAGE/LANE

The derivation of the name of this ancient public footpath which runs between Humber Street and Hope Street is unknown. However, it could derive from the name of local resident Joseph Grant who was born in 1807. A fisherman and oyster dealer, he opened an oyster-booth on the Cleethorpes beach and also hired out bathing machines to hardy Victorian bathers. He prospered and in 1856 bought land in the location of where we now have Grannies Passage; and this is where his connection with the passage comes in. Like many Cleethorpes Victorian residents Grant had a nickname – which was 'Granny' – based on his surname. So probably Grannies Passage was named after this local landowner, Joseph 'Granny' Grant. He died in 1861 but the family connection with the area remained. His widow, Mary, outlived him by 33 years and in 1878 a Mrs Grant was keeping a 'booth' at the bottom end of South Street. There is still a single-storey building there, only a few yards from Grannies Passage.

32 *Grannies Passage public right of way between Humber Street and Hope Street, in 2008.*

GRANT STREET

This street was built on land belonging to Sidney Sussex College. The stretch between the High Street and the present Station Road was actually part of Station Road from 1878 when the Cleethorpes Local Board decided that the road leading from the *Victoria Hotel* (currently *O'Neill's Irish Pub*) to the railway station should be called Station Road. The full length of what became Grant Street was laid out by 1887 but major development did not begin to take place until 1895 when the College set out building sites in the street. By that time it had acquired its present name. Being on College land one would expect it to have a name associated with the College but there are no known contenders. The name remains a puzzle.

GREETHAM'S LANE

This lane in Old Clee led to the farm run by Robert Greetham; and run later by his three sons Harry, William and Harold.

GRIMSBY ROAD

In 1846, the Enclosure Award Commissioner set out a 'Public Carriage Road of the breadth of thirty feet called the Grimsby Road'. It was described as running from the bottom of Isaac's Hill 'and extending in a Northwardly direction ... until it enters the Parish of Great Grimsby near the Ropery' – this was the point where the present Humber Street runs.

33   *Greetham's Lane in Old Clee, and a much-altered cottage in 1987.*

The road would presumably have been an improvement on the previous road or track to Grimsby which, going by the 1824 Ordnance Survey map and the pre-enclosure map, may have been further to the east and running across the Horse Course Marsh. The Enclosure Commissioner appears to have set Grimsby Road further west following the boundary between the Horse Course Marsh to the east and the Clee Common Marsh and Clee Cow Marsh to the west.

34   *Grimsby Road c.1906, looking towards Grimsby. The house on the left-hand side with the tall chimney is on the corner of Tiverton Street.*

36 *Beaconthorpe Methodist Church and the corner of the old gas showrooms, Grimsby Road. Both buildings date from 1914 and are seen here in 1991.*

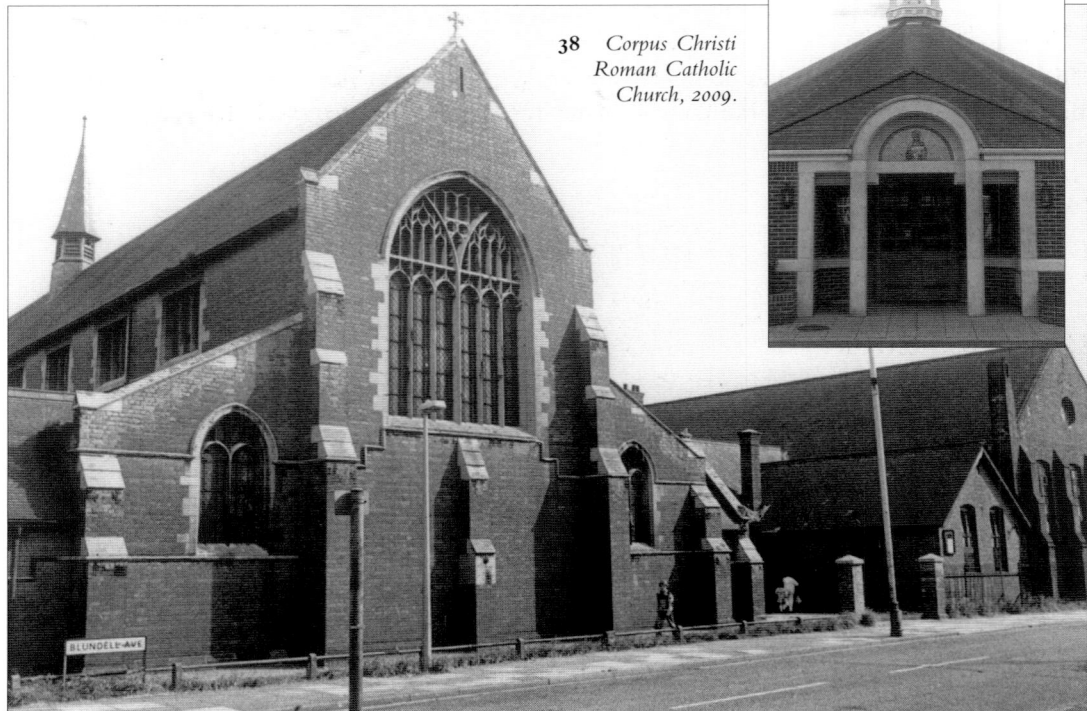

35 *The electricity showrooms on the corner of Grimsby Road and Prince's Road which were opened in 1937.*

38 *Corpus Christi Roman Catholic Church, 2009.*

37 *St. Aidan's Church, consecrated 1906. The church building on the right was used as a hospital during the First World War and its site now contains a block of flats.*

The road contains several historically-important buildings. These include St Aidan's Church (1906), the Constitutional Club (1911-12), the Beaconthorpe Methodist Church (1914), the old gas showrooms (1914), the old art-deco electricity showrooms (1937) and the Memorial Hall (1960). This latter building, on the corner of Clee Road, commemorates the sacrifice of Cleethorpes people who were killed in the Second World War. The plaques in the Memorial Hall entrance record 368 fatalities of the conflict. These consist of 112 crew members of minesweepers and fishing vessels, 83 soldiers, 60 sailors, 53 airmen, 35 civilians and 25 merchant seamen.

GROSVENOR COURT *see* BEDFORD ROAD

GUNBY PLACE

The names on a large part of the Cleethorpes Council's extensive Beacon Hill estate are on the theme of castles, stately homes, abbeys, etc. Gunby Hall, between Spilsby and Skegness, was built in 1700 for Sir William Massingberd and was reputed to be Tennyson's 'haunt of ancient peace'.

HAIGH STREET

William Haigh owned land in Cleethorpes in 1846. This street was set out sometime after 1887 and completely built up by 1906. Nowadays it gives pedestrian access to the Brooklands Avenue enclave.

HALTON PLACE

Roads in this area are named after palaces, castles, etc. There is a Halton Castle in Northumberland.

HAMPTON COURT

Roads in this area are named after palaces, castles, etc. Hampton Court Palace by the River Thames was built during 1515-25 for Cardinal Wolsey. He later handed it over to Henry VIII. It ceased to be a royal residence in the 18th century and was the first royal palace to be opened to the public, in early Victorian times.

HARDY'S ROAD

This road has had a number of name changes. In 1846, it was referred to as the Thrunscoe Occupation Road; an occupation road was one that gave access to farm land. It ran along the line of the present Hardy's Road but ended at the Thrunscoe Farm, some yards short of the present Chichester Road. It later became known as Thrunscoe Lane. Then, in 1900, J. Hardy rented the Thrunscoe Farm (which became known as Hardy's Farm) from Sidney Sussex College at an annual rent of £270. The road was still Thrunscoe Lane when,

39 *Thrunscoe development, 1963. To the right, Hardy's Road comes from the top with Cromwell Road and Daggett Road leading off. To the left are Minshull Road and Weekes Road, running between Aldrich Road and Pearson Road. Hardy's Farm buildings are in the lower right-hand corner, now the site of the Signhills Schools.*

in 1960, the College started to develop the area for housing. The Cleethorpes Council then suggested changing the name to Hardy's Lane to avoid confusion with the present Thrunscoe Road. Two years later the name Hardy's Road was in use when new freehold semi-detached 3-bedroomed houses were being advertised for sale there at £3,050 'ready for occupation now'. The Hardy's Farm buildings were located on the site of the present Signhills Schools.

HARRINGTON STREET

Sir John Harington (note only one 'r') later the 1st Lord Harington of Exton, was a nephew of Lady Frances Sidney Sussex the founder of Sidney Sussex College. He was the joint executor of her will, carried out her wishes in setting up the new college and 'may justly be regarded, equally with Lady Sidney as a Founder of the College'. He contributed at least as much money to the College as was bequeathed by her and also negotiated for the site and supervised the three-year building operation. He endowed the college with his Lincolnshire manor of Saleby.

He was also made guardian of James I's daughter Princess Elizabeth 'but she proved to be so extravagant that the money granted for her maintenance (£1,500 a year) was hopelessly insufficient, and Harington had to dip deeply into his own pocket.' To relieve his difficulties, he was granted the sole right to coin brass farthings for three years and these coins were accordingly known as 'Haringtons'. He prevented the abduction of the princess by gunpowder plotters in 1605. He died in 1613. He is sometimes confused with his cousin, Sir John Harington (1561-1612) who was a godson of Queen Elizabeth and who is credited with being one of the inventors and popularisers of the flush toilet. The development of Harrington Street started in 1891 when application was made to build 19 houses there.

HART STREET

Sir John Hart was an Alderman (1620-40) and Lord Mayor of London. He left a bequest to Sidney Sussex College to buy an estate to help the support of the Master, a Lecturer in Greek, two Fellows and two poor Scholars – with preference for men from Coxwold School. This school in North Yorkshire was founded by Sir John in 1600.

HAVERSTOE PARK

Originally allotments on Sidney Sussex College land, the park was opened in 1974 and visited by the Queen on 12 July 1977 to open the park's Jubilee Sensory Garden. For an explanation of the park's name *see* the entry for Haverstoe Place.

HAVERSTOE PLACE

Most of the roads in this corner of the town have names with a local or Lincolnshire theme. Haverstoe was the name of the wapentake in which Cleethorpes lay. Wapentakes were the administrative divisions of Lincolnshire at the time of Domesday Book in 1086.

HAWTHORN AVENUE

Most of this road was built on the land of Chapman's brick works. In 1926, the Cleethorpes Council approved plans for part of the road to be built off Suggitt's Lane. This would meet up with the rest of the road which was to be built off Chapman Road. The two parts can be distinguished by the change in building styles about half-way along. The complete new road was adopted by the Council in 1928.

HEWITT'S AVENUE *and* HEWITT'S CIRCUS

Named after the Grimsby brewer T.W.G. Hewitt. He gave land so that major road improvements could be carried out to cater for the increasing traffic which was using the junction of Taylor's Avenue, Humberston Road and what became Hewitt's Avenue. In January 1930, work began on getting tenders for the work to be carried out on improving the junction; Mr Hewitt died the following May aged 73. Two months later the work was completed and public benches were provided. The traffic island in the centre of the Circus was not part of the initial scheme but work began on it in 1937.

HEY STREET *see* SHERBURN STREET

HEYTHROP ROAD

Roads in this part of the resort are named after fox hunts in various parts of the country. This one is named after the Heythrop Hunt in Oxfordshire and Gloucestershire.

40 Cliff Hotel *on High Cliff, in the 1950s. Now the site of 'The Point' high-rise flats.*

HIGH CLIFF

The highest point of the Cleethorpes cliffs, now about fifty feet above sea level. It was once higher but its height was reduced during the seafront improvements carried out in the early 1880s. The top of the Ross Castle is said to mark its original height. High Cliff was the site of Cliff House which was built by Richard Chapman. In 1850, he advertised sea bathing, for which he had: 'Eight Machines for Sea-Bathing in constant readiness.' He also had: 'Rooms for the accommodation of Two Private Families, with full sea views [and] Two separate Rooms for the Sale of Ale and Porter, to be consumed on the Premises, or sold to Families at the lowest profit.' In 1853, the building was extended to form the *Cliff Hotel*. The building went through several transformations and was demolished in 2004, to be replaced by the high-rise apartment block named 'The Point'.

41 *A sunny afternoon at the decorative shelter which stood on High Cliff opposite Sea View Street, seen here in the 1950s.*

HIGH CLIFF ROAD

This is an obvious name for the road, which leads up
to High Cliff from the Kingsway. However, the lower
part used to be known as Fisherman's Road (*see* the
entry under that name). Building land there was sold
by Sidney Sussex College in the 1840s to help meet
its costs under the Enclosure Award. The present High
Cliff Terrace was then built north of Humber Street. It
was described in 1850 as 'a splendid pile of buildings,
containing seven houses fit for the reception of large
families; they command a full view of the sea, from which
they are only a few yards distant'. In the same year all
the houses were available as lodgings for visitors. Another
house was added to the terrace at its southern end.

HIGH STREET

The name High Street is usually interpreted as meaning
the main or principal street of a town. But that is debatable
in the case of this street. It is certainly an old thoroughfare
and provided the most direct route to the seafront for
visitors arriving by road from Grimsby. But it lay on the
northern fringe of the town and was recorded as North
Street in 1858. The 1850s and 1860s saw increased building
in the expanding town and the present name appears to
have come into general use by 1872.

42 *Cliff Cottage which was tucked-in behind the*
Nottingham Hotel *and seen here in 1994. Since
then it has been cleared for 'The Point' high-rise
flats on High Cliff.*

HIGH THORPE *and* HIGHTHORPE CRESCENT

High Thorpe was an alternative
name for Itterby and has been
recorded in use in 1749, 1851, 1866
and 1871. It was an understandable
name in view of the location of
Itterby on the highest part of the
seafront. The name has since been
used to create the name of the
modern Highthorpe Crescent, off
Middlethorpe Road.

43 *High Street in Edwardian days,
looking in the direction of Isaac's Hill.*

44 *High Street looking east, 1923. St Peter's Avenue leads off on the right and the* Leeds Arms Hotel *is in the centre of the picture.*

45 *The Coliseum in the High Street. Built as a cinema in 1920 it has had several other uses, including snooker hall, indoor market and Woolworths store. It is currently a nightclub.*

HIGHGATE

Highgate as we know it today comprises two parts. Work on constructing the modern stretch, between Thrunscoe Road and Trinity Road began in the first half of the 1930s. The other part, between Cambridge Street and Thrunscoe Road, is considerably older. In the older part, south of Oxford Street is a row of houses with a stone bearing the name 'Ludborough Terrace A.D. 1861'.

The usual interpretation of the name Highgate is 'Main Street'. This is from 'High' meaning main or principal and 'gate' from the Scandinavian 'gata' meaning street. However, it was an unlikely main street and in 1846 was merely included as part of the Thrunscoe Road. Another interpretation of the name was suggested by a local historian C.E. Watson in 1901. He suggested the meaning 'High-garth, the paddock on the rising ground'. In support of his suggestion, it is noteworthy that in 1734 the name 'Hygate' appeared in a Sidney Sussex College land document and the street actually starts in the vicinity of some old inclosures of College land. Therefore, it seems that the street could have been named after a piece of College land. But all we can safely say is that the jury is still out on the origin of this street name.

46 *The* Leeds Arms Hotel *on the corner of High Street and Cross Street. Later the site of a Woolworths store. Photograph taken from Cross Street, c.1955.*

HILTON COURT *see* BEDFORD ROAD

HINKLER STREET

Bert Hinkler, a famous Australian flier, made the first solo flight from Britain to Australia in 1928, taking 15 days. In 1933 builders Taylor & Coulbeck asked for this recently constructed cul-de-sac to be named Hinkler Street. The request was agreed by the Cleethorpes Council, which adopted the road in 1934. *See also* Mollison Street.

HOOLE *see* OOLE

HOPE STREET *see* NEW BRIGHTON

HORSE COURSE MARSH *see* RACE GROUND

HOWLETT ROAD

Richard Howlett was made a Fellow of Sidney Sussex College in 1610. He was appointed tutor to the young Oliver Cromwell when the latter became a student at the College in 1616. Therefore, it cannot be a coincidence that this road leads off Cromwell Road.

HUMBER STREET

An obvious name in view of the vistas of the Humber from the end of the street. It is a very old street, being shown on the Enclosure Award plan of 1846, and had acquired the name Humber Street by 1861. A stone, inscribed 'Itterby Terrace 1875' is above the archway in Humber Street, opposite the entrance to Grannies Passage. In 1891, the terrace also had a name plate bearing the name Bancroft Court which, in that year, was removed at the residents' request. The street was largely built-up by 1887.

One resident stated in 1991 that some of the houses in the street were known as 'apprentice houses' because there was a solid wall separating the front from the back of the houses in order to protect the bosses' daughters from the attention of (lustful?) apprentices – who lived in the rear part of the house. It makes a good story but were they just straightforward back-to-back houses?

A different form of excitement came to the street in 1882 when an electric power plant was situated there in order to provide street lighting in the town. Three lamps were installed initially – one near the *Cliff Hotel* and the others in Sea View Street and at the corner of Mill Road. They were switched on, on 14 September 1882, and the service was later extended to other streets. Unfortunately, the overhead bare wires which carried the current on poles were blown together on windy nights, causing short circuits and fusing – and providing what were called 'pyrotechnic displays'. Not surprisingly, the three-year contract was not renewed for this enterprising scheme with its unexpected visually-exciting enhancements.

HUMBERSTON ROAD

This was given its name because it led from the village of Clee to the village of Humberston. It was described in the 1846 Enclosure Award as a 'Public Carriage Road of the breadth of Thirty feet called the "Humberstone [*sic*] Road" commencing at the Guide Post [Love Lane Corner] near the village of Clee and extending in a Southwardly direction … to a certain bridge [at Buck Beck] where it enters the parish of Humberston'. The road was widened in the 1930s.

⋊⟹⟸⋉

IMPERIAL AVENUE

Named after the nearby *Imperial Hotel*, the name of which is not surprising. The hotel was being built at the end of the 19th century; a time of imperial grandeur when the British Empire was ruled by Queen Victoria, whose imperial status had been achieved when she was made Empress of India in 1876. In 1930, the Cleethorpes Council agreed to plans by builders Taylor & Coulbeck to lay out land in front of the Blundell Park Football Ground and build there 47 houses which would comprise Imperial Avenue and Constitutional Avenue. By 1935, all the houses in the two avenues were occupied. *See also* the entry for Blundell Park Football Ground.

IRBY COURT

Most of the roads in this corner of the town have names with a local or Lincolnshire theme. Irby-upon-Humber lies only a few miles from Cleethorpes and produced a hero of the Civil War. Royalist Edward Lake received 16 wounds in a battle but with one hand useless fought on with his sword in the other. Imprisoned by the Parliamentarians, he escaped, lived to see the Restoration of the monarchy in 1660, was made a baronet and was buried in Lincoln Cathedral.

47 *Imperial Avenue, with spectators making their way to the Blundell Park Football Ground, 1950s.*

48 *The residential home on Isaac's Hill in 2008. It was built as a Technical Institute during 1901 and later housed the public library until 1984.*

ISAAC'S HILL

One of Cleethorpes' best-known street names has no definite origin. In 1901, local historian C.E. Watson wrote that the name was 'lacking a reasonable derivation'. In 1953, another historian, Dr Frank Baker, suggested that it may derive from the name of the two Isaac Chapmans, father and son, who formerly owned the *Cross Keys* public house. This stood at the top of the hill, where the car park is now sited just off the High Street.

Another contender is a doctor called Isaac Dan who apparently lived at the top of the hill in the early 1800s. It was reported that he became deranged and spent much time in a tree-house in his garden. Whilst there, 'He sermonised from this high place and urged Neptune to rise and drown the populace. Such direful and doleful raving became a spectacle and led to the cognomen [nickname] Isaac's Hill. The doctor eventually died of exposure!'

The street name is absent from local documents. In the 1846 Enclosure Award, the hill is just referred to as part of the road to Weelsby. Neither was the name used in the local directories and guides of 1850 and 1858. The first official use of the name 'Isaac's Hill' discovered so far is in the 1895 Cleethorpes Council minutes. Accordingly, it may be surmised that a popular nickname was adopted as the official name in the 1890s when the road began to be built-up.

The incline of the hill has been altered over the years. The horse-drawn tram service from Grimsby was extended as far as Poplar Road in 1887 but stopped short of the steep Isaac's Hill. Then, in 1898, the gradient of the hill was 'eased' and the service extended to

Albert Road. The Cleethorpes Town Hall was originally going to be built at the top of Isaac's Hill but instead the site was used for a Technical Institute, which was built in 1902. The building later housed the public library and is now a residential home.

ITTERBY *and* ITTERBY CRESCENT

The hamlet of Itterby was one of the three thorpes of Clee, the others being Oole and Thrunscoe, which all eventually grew together to form modern Cleethorpes. The thorpes were outlying hamlets of the parent village of Clee (now known as Old Clee). Itterby was centred on Sea View Street and Wardall Street. The name comes from the Scandinavian words 'ytri' and 'by', which taken together mean an outer farm or settlement. It was listed in 1086 in Domesday Book, under the spelling Itrebi. It has also been known locally as Middle Thorpe, Upper Thorpe, High Thorpe and Far Cleethorpes. There were 18 families living there in 1563. The name has been revived in the modern Itterby Crescent. *See also* the entry for Cleethorpes.

ITTERBY ROAD

The road was given this name because it led from Oole to Itterby. It was described in the 1846 Enclosure Award as a 'Public Carriage Road of the breadth of Thirty feet called the "Itterby Road" commencing at the *Dolphin Hotel* … and extending … to the Town Street [Humber Street]'. It is now known as Alexandra Road except for the southerly portion which is the northern stretch of High Cliff Road.

　　　In the mid-19th century the road contained Mantell's Buildings. These were built on land belonging to the Rev. Edward Reginald Mantell. Yarra Road was built through the centre of his land, which also now includes the Cleethorpes Library. In Dobson's 1850 directory Mantell's Buildings were described as 'the splendid buildings lately erected by the Rev. E.R. Mantell, Vicar of Louth, which have the Pleasure Ground and sea in front; and the National School in the rear of the premises'. Mantell had been vicar of Tetney in 1836. In 1850, two lodging-house keepers were listed at Mantell's Buildings, Mary Healas and David Swanson.

ITTERBY TERRACE *see* HUMBER STREET

⋅⊶⊙⊷⋅

JAMES BUILDINGS, JAMES SQUARE, JAMES STREET

These names were all used in the close-packed cluster of small houses which were once on what is now the Wardall Street car park. They were probably named after James Appleyard, who in 1850 was listed as a lodging-house keeper in James Square. Listed as a lodging-house keeper in James Street at the same time was Levi Stephenson. He was a well-known local fisherman, boatman and all-round 'character' who was featured on the coat of arms adopted by the Cleethorpes Borough Council in 1936. *See also* Wardall Street.

Jenner Place

David Jenner was a tutor to William Wollaston at Sidney Sussex College when the latter was admitted in 1674. In 1931, the Cleethorpes Council agreed to the name for this new cul-de-sac. *See* the entry for Wollaston Road for information on William Wollaston.

Johnson Street

Dr James Johnson was Master of Sidney Sussex College, 1688-1704. He was regarded as a 'thoroughly jovial person' and bequeathed £1,200 to the College for the purchase of church livings. He also left the college several estates, including one at Cherry-Hinton near Cambridge, for the maintenance of three or four 'poor widdows of Clergymen who have been of the College' and for the support of orphans of clergymen. Unfortunately, his will had not been signed by witnesses so the estates went to his heir-in-law. The College was later able to purchase the Cherry-Hinton estate.

⊶═◦═⊷

Kathleen Avenue

The Cleethorpes Council agreed to the name of this street of semi-detached houses in 1930, and adopted the road in the same year.

Kenilworth Road

Roads in this area are named after palaces, castles, etc. Kenilworth Castle in Warwickshire was one of the strongholds of the dissidents in the civil war of the 1260s.

Kew Road

In default of information to the contrary, we have to assume the name is taken from the famous Kew Gardens. In 1906, the land which now contains the road and Elm Gardens, was a two-acre field. By 1923, the field was fully built up with the extended Bowling Lane running between Kew Road and Elm Gardens.

Kingsway *and* King's Parade Promenade

This road and the parallel promenade were opened on 12 July 1906 by Lady Henderson, the wife of the chairman of the Great Central Railway Company. The road and promenade replaced the badly-eroded cliffs of Sea Bank Road and were presumably named after the new king, Edward VII, who had succeeded to the throne in 1901. Constructed by the Cleethorpes Council, they were a popular visitor attraction and also enabled the southern part of the resort to be opened up for housing and recreation. In 1920, the magistrates fined a cyclist one pound for cycling on the King's Parade and also fined motorists from Goole a pound for motoring on the Parade.

The Kingsway once hosted two popular dance venues. One was the Café Dansant which opened in 1914 as the Kingsway Pavilion and was renamed the Café Dansant in

49 *The southern end of the busy King's Parade, probably in 1914, with a band shelter in the background.*

the early 1930s. It was closed in 1966 and then demolished. The other was the Winter Gardens which opened as the Olympia in the early 1930s. It was refurbished and re-opened in 1947 as the Winter Gardens but was demolished in 2007.

Kingsway Gardens

The space occupied by these delightful gardens which stretch the length of the Kingsway was meant to accommodate tramlines. At the time the Kingsway was being constructed, the tram terminus was at Brighton Street. The intention was that when the road was completed the trams would run to the end of the Kingsway. However, negotiations between the Cleethorpes Council and the tramway company broke down so the land was quickly made into the Kingsway Gardens.

50 *An Edwardian view of the Kingsway, taken from opposite Brighton Street. The houses on the near-right are now retail premises and a block of flats.*

The most famous 'resident' in the gardens is The Boy with the Leaking Boot statuette. He was donated to the resort in 1918 by John Carlbom who came to England in the 1890s from his native Sweden, founded a successful company in Grimsby and became the Swedish Vice-Consul. He lived in Cleethorpes until 1910 and presented the statuette in appreciation of the time he and his family had lived in the resort. The 'Boy' is subject to vandalism but hopefully will continue to reside in the Kingsway Gardens.

KING'S ROAD

This road would have been given its name because it is a continuation of the Kingsway. The Kingsway was opened in 1906 and ended where the Leisure Centre now lies. Beyond that point was the golf course and salt marshes which were prone to tidal flooding. There was just a track which led to the parish boundary at the Buck Beck. A bridge took the track over the beck and it continued in the Humberston parish. Work on improving the track was held up because of differences between the Cleethorpes Council and Sidney Sussex

51 *Kingsway Gardens and the Boy with the Leaking Boot, c.1920s.*

52 *The Café Dansant on the Kingsway. It was closed in 1966 and demolished.*

53 *The Winter Gardens on the Kingsway. It was demolished in 2007.*

54   *An Edwardian view of the King's Parade promenade, the Kingsway Gardens and the Kingsway. The premises on the extreme right are now the* Kingsway Hotel *on the corner of Queen's Parade.*

College, which owned the land in that vicinity. However, in 1938, the Council was able to approve plans for a King's Road dual carriageway, which was completed in May 1939, extending just over a thousand feet from the Kingsway. The central islands were planted with flowers. 1985 saw the beginning of major improvements to the extended road and the building of a new bridge over the Buck Beck.

KNOLL STREET

Named because of its location on high ground, a knoll being a hillock or the top of a hill. Knoll Street was originally only the short stretch from Alexandra Road to New Street. It was given its name in 1878 by the Cleethorpes Local Health Board. The extension of the street to Cambridge Street was completed and dedicated in 1898.

55   *Ye Olde Barn Teahouse which was on the corner of Brighton Street and High Cliff Road. The building is now a wine bar and public house.*

LADY FRANCES CRESCENT

The land for the streets in this area was originally owned by Sidney Sussex College. Accordingly, this crescent is named after the founder of the College, Lady Frances Sidney, Countess of Sussex (1531-89). She was buried in Westminster Abbey where her substantial monument may still be seen. In her will she left £5,000 for the foundation of the College, which opened in 1596. The supervisor of her will was John Whitgift, who had been born in Grimsby and was Queen Elizabeth's Archbishop of Canterbury.

LANGLEY PLACE

The name derives from Abbots Langley in Hertfordshire where Sidney Sussex College owned an estate. *See* the entry for Combe Street for more information.

LESTRANGE STREET

Sir Roger L'Estrange (1616-1704) was a student at Sidney Sussex College. He has been described as the first great political journalist. He was on the Royalist side during the Civil War. A College historian has commented that: 'In early life he was quite the Don Quixote of the Royalist party. His attempt to recover Kings Lynn for the King in 1644 is one of the funniest episodes in English history. In this wild enterprise he was taken prisoner by the Parliament and condemned to death. Afterwards he was reprieved and kept in Newgate [prison] till 1648. He gave up soldiering and took to literature; he was a most zealous and successful pamphleteer in favour of Monarchy and the Church'.
Application to start building in the street, with 49 houses, was made in 1914. The section of the street between Grimsby Road and Brereton Avenue was completed by 1931. Further building took place during the 1930s in the part beyond Brereton Avenue, which was adopted by the Cleethorpes Council in 1936.

LEWIS ROAD

This short cul-de-sac of semi-detached houses was completed by 1931. The Lewis family lived in the 1880s in Poplar Villas, opposite the land where Lewis Road lies. As they were market gardeners we may assume that the road was built on their land.

LINCOLN ROAD

Streets in this small corner assembly are named after prestigious places, such as Lincoln, which have castles, cathedrals or stately homes.

LINDSEY ROAD *see* SHERBURN STREET

LINDUM ROAD

Lindum Colonia was the Roman name for Lincoln, so it is an appropriate name for a Lincolnshire road. The land in this area of Thrunscoe was part of White's Farm. The farm was purchased by the Boulevard Estate Company who began to develop the land for housing. The construction of the Kingsway opened up the area for building, and plans for the layout of Lindum Road, Seacroft Road and Signhills Avenue were approved in 1912. However, building did not take place until after the First World War, with much building activity taking place from the early 1920s.

LINKS ROAD

As this road is close to the golf course, or 'links', the name is self-explanatory.

LOVE LANE CORNER

A charming, but unlikely, name for a busy road junction. However, the name is an old one and arose out of the footpath which was known as Love Lane and which still leads off the junction here (by the two old cottages) and which once had a kissing gate. The junction is an ancient meeting point of three roads and its name was given official recognition by the Cleethorpes Council in 1935. Two years later the Council's Highways Committee recommended that the present roundabout should be constructed.

LOVEDEN COURT

One of the cluster of roads south of Chichester Road with names on a Lincolnshire theme. Loveden is the name of one of the ancient wapentakes which were the administrative divisions of Lincolnshire at the time of Domesday Book in 1086. Loveden wapentake was located south of Lincoln, bordering on Nottinghamshire and Leicestershire.

LOVETT STREET

Thomas Lovett was a student at Sidney Sussex College and also a College benefactor. He left £2,000 in 1776 for the support of sons of clergymen at the College. Preference was to be given to scholars from Grantham and Oakham schools. The legacy was used to purchase land at Melton Mowbray, which produced about £80 annually.

LOW THORPE *see* OOLE

LUDBOROUGH WAY

One of the cluster of roads south of Chichester Road with names on a Lincolnshire theme. Named after the nearby village of Ludborough, which was in the Ludborough wapentake. Wapentakes were the administrative divisions of Lincolnshire at the time of Domesday Book in 1086.

⋅⊷≡◉⊂⊷⋅

MACHRAY PLACE

Dr Robert Machray (1831-1904) was a student at Sidney Sussex College and was appointed College Dean in 1858. Subsequently, he had an illustrious church career in Canada. In 1865, he was appointed the first Archbishop of Rupertsland (now Manitoba) and 'For

May Street

Thomas May (1595-1650) became a student at Sidney Sussex College in 1609. He had 'an imperfection in his speech, which was a great mortification to him'. He wrote plays and composed several narrative poems for Charles I. His poetry was praised by the poet Ben Jonson. He was disappointed at not becoming Poet Laureate on Jonson's death and transferred his allegiance from the King to Parliament. He was very energetic in supporting Parliament up to the time of his death, which was said to have been caused by 'tying his nightcap too close under his fat chin and cheeks'. He was buried in Westminster Abbey but in 1660 at the Restoration of the monarchy his body was dug up and thrown into a pit in St Margaret's Churchyard – and his monument was demolished.

Mayfair Court *see* Bedford Road

Megs Island

Megs Island is a piece of local slang, meaning that part of Cleethorpes which is 'above' Isaac's Hill – i.e. the older, resort part of Cleethorpes. In turn, it has produced the term Meggie (or Meggy), meaning a native of Cleethorpes who was born and bred in old Cleethorpes 'above the hill'. The derivation of Megs Island is unknown. Local historian C.E. Watson wrote in 1901 that 'the slang name for Cleethorpes is, and apparently has been for centuries, Megs Island'. He then suggested that this came from 'Mag Heiland' derived from old English and Teutonic words meaning Big Hill. He suggested that this was: 'a reasonable name enough for the rising ground on which Cleethorpes stands, in a district destitute of hills for miles around'. Others think that the name came from the price of a tram or railway ticket from Grimsby to Cleethorpes, i.e. a 'meg' or halfpenny. However, this is unlikely as the name seems to have been in use long before the days of trams and trains.

Meridian Embankment

Originally called the Marine Embankment, it was recently given its present name because the Greenwich Meridian of longitude passes through it. It is a sea bank, footpath and cycle way which stretches for just over a mile along the seafront; from the Boating Lake paddock to the car park at the end of Anthony's Bank Road. It was opened on 6 September 1930 and enclosed 115 acres of fitty land (salt marsh) which was subject to tidal flooding. The land was then drained and improved. It now contains the

58 *Meridian Line on the Meridian Embankment, 1993. The metal Meridian strip was placed there in 1933.*

Pleasure Island amusement park and the Meridian Park events arena. The Greenwich Meridian metal strip was placed on the embankment by the Cleethorpes Council in 1933.

MIDDLE STREET *see* WARDALL STREET

MIDDLE THORPE *see* ITTERBY

MIDDLETHORPE ROAD

Most of the roads in this corner of the town have names with a local or Lincolnshire theme. Middle Thorpe was one of the alternative names used in the past for the thorpe of Itterby. Also, this modern road is on land which in the 1740s was called the 'Thorpes Middle or Great Field'. This was one of the large open fields which were farmed communally at that time. A simple inversion of 'Thorpes Middle' (and deleting an 's') also gives us Middlethorpe – all part of the fascination of street names. Not for the faint-hearted house hunter is the news that semi-detached houses were being built on the road in 1968 for sale at £3,350 freehold.

MILL

Several street names include the word 'Mill'. Accordingly, in order to avoid confusion we should note that Cleethorpes had two windmills. The street names Mill Road, Millers Road and Mill Hill Crescent refer to the four-sailed mill which was sited near the top of Mill Road. The names Mill Place, Mill Garth and Millers Garth refer to the five-sailed mill which was sited in Mill Place, off the Market Place.

MILL GARTH

Most of the roads in this corner of the town have names with a local or Lincolnshire theme. Mill Garth is probably based on the old local name Millers Garth, which later became Mill Place.

MILL HILL CRESCENT

Given this name because of its nearness to the top of Mill Road and the site of the old windmill – *see* the entry on Mill Road for further information on the mill. In 1931, the Cleethorpes Council approved plans from builder J. Would for 16 houses in this cul-de-sac. The name was approved by the Council in 1932.

MILL PLACE

In 1846 this location was referred to as Millers Garth and later became known as Mill Place.

59 *Queen Victoria stone, dated 1891, set in house wall in Mill Place.*

In the mid-19th century the five-sailed windmill there, and the adjoining bakehouse, were run by Methodists John and Charles Loft. The mill was still operating in 1887 when there was open land in front of it. But by 1906, this open land had been built-up with Dolphin Street and seafront property. This would impede the wind currents needed by the mill. Accordingly, it is uncertain whether the mill was still operating then. A millstone on the western side of the road, once marked the location of the mill, but the site has been built over in recent times.

MILL ROAD

Originally known as Crow Hill, it was known as Mill Road by 1846 but the name Crow Hill also continued to be used. The four-sailed windmill was near the top of the hill and was on land allocated to John Nicholson in 1846; he and Thomas Frankish worked the mill. In 1853, Frankish 'miller and baker' issued a business card stating that 'Having engaged an Experienced Man in the Baking and Confectionary department, he hopes to merit public patronage. N.B. Families supplied with hot bread every morning at their houses'. The mill was bought by Bratley's of Brigg in 1871. By 1931 it was 'disused' but was still there in a derelict condition in the 1950s. *See also* the entry for Crow Hill Avenue.

At the beginning of the 1900s, Alderman William Grant had a large villa built at the very top of the

60 *Pair of semi-detached villas in Mill Road, built 1887, seen here in 1985.*

61 *Detached villa in Mill Road, 2008.*

62 *Remains of the Mill Road windmill, 1950s. The resort's large 1931 gas holder (demolished 1967) is in the distance on the extreme right.*

road. It was called 'The Mount'. After his death in 1926 it was purchased by Alderman Mrs Ada Croft Baker and donated for use as a maternity home. Accordingly, what most residents refer to as the Croft Baker Maternity Home was opened in 1929. It was closed in 1984 and the site is now 'The Mount' development of bungalows and flats.

The street has an interesting selection of name stones for individual houses and for terraces such as Dowsing Crescent, John's Terrace and Lyndhurst Terrace. There is also the charming row of cottages bearing the inscription 'King George V Coronation Homes 1911'.

63 *'The Mount' in Mill Road built c.1900. In 1929 it became the Croft Baker Maternity Home.*

MILLERS GARTH *see* MILL PLACE

MILLERS ROAD *see* BEACON AVENUE

MINSHULL ROAD

Built on Sidney Sussex College land in the early 1960s, the name of this quiet residential road takes us back to some far-from-quiet times. In 1643, the 11 Fellows of Sidney Sussex College had to elect a new Master. There were two candidates for election: Richard Minshull, a Parliamentarian, and Herbert Thorndike, a Royalist. The latter was likely to be elected by a majority of one vote. The election was preceded by Holy Communion in the College chapel but Parliamentarian soldiers rushed in during the service, seized one of

Thorndike's supporters and hustled him off to jail. Four more of Thorndike's supporters left the chapel in protest – and Minshull was elected by five votes. Thorndike's supporters complained but to no avail. So Dr Richard Minshull remained Master of Sidney Sussex College from 1643 to 1686. College discipline amongst students was poor during his Mastership and drunkenness was a common offence. Five students were expelled from the College: one for threatening with sword and pistol, another for breaking the Dean's windows, and three more for attempting to burgle the Master's Lodge.

MOLLISON STREET

Along with nearby Hinkler Street, this road takes us back to the days of famous air pioneers. In 1930, Amy Johnson (1903-41), later Mollison, was the first woman to fly solo from England to Australia; the flight took 19½ days. She made other record flights with J.A. Mollison, whom she married in 1932. Mollison, a former RAF pilot, was himself a record-breaking flyer. He flew from Australia to England in 10 days in 1931. The couple divorced in 1938. Amy died whilst ferrying aircraft during the Second World War. The street was adopted by the Cleethorpes Council in 1934.

MONTAGUE STREET

Dr James Montagu (note no final 'e') was the first Master of Sidney Sussex College, 1596-1608. He was the great-nephew of the College founder, Lady Frances Sidney Sussex. He was also the nephew of Sir John Harington (*see* the entry for Harrington Street). It was said that 'the Montagu clan made the College their own and supported it by benefactions'. Dr Montagu became Bishop of Bath and Wells in 1608 and then Bishop of Winchester in 1616. He died in 1618; see his grand tomb in Bath Abbey. He bequeathed a perpetual annuity of £20 to the College. His brother Edward gave the College a thousand-year lease of 45 acres in Burwash in Sussex. The Montagu connection with the College continued through the 17th century, nine nephews and seven great-nephews of the first Master becoming students at the College.

MONTGOMERY ROAD *and* NORMANDY ROAD

Two post-Second World War roads which commemorate Field Marshal Bernard Montgomery and the Normandy invasion.

MUCKY LANE *see* CORONATION ROAD

NEAR CLEETHORPES *see* OOLE

NEPTUNE STREET

It has been written that this street was named after a Russian ship which was bought by Robert Keetley in 1852. Neptune Terrace was listed in a directory of 1871 and part of the

street was built by 1887. The coastguard station was located here in 1890. In 1920, seven houses in the street still had old unhygienic 'privies' which the Cleethorpes Council had been getting rid of in the resort since the beginning of the century – accordingly it told the house owners to get them converted to water closets.

NEVILLE STREET

Sidney Sussex College had two members of this name who were worthy of commemoration. Dr Thomas Nevile (note only one 'l') was Master of Trinity College. He was instrumental in helping the creation of Sidney Sussex College by providing the site where the College was built in 1595. Leaping forward about three hundred years, we then have Francis Henry Neville (1847-1915) who entered Sidney Sussex College in 1867. He held the important post of College Bursar during 1890-96 and also managed the College laboratory from 1880 for 28 years. He was well-known as a scientific researcher, particularly into metallurgy. He also trained many significant scientists. It has been suggested (with tongue firmly in cheek) that one of these was the fictional Sherlock Holmes, who had supposedly studied at Cambridge. Accordingly, he was tentatively identified by the crime writer Dorothy L. Sayers as one T.S. Holmes who entered the college in 1871! Leaving fiction behind, we note prosaically that Neville Street was built-up during 1907-8 when 110 houses were planned to be built.

NEW BRIGHTON

This was the name acquired by that cluster of streets opposite the Brighton Street slipway. In the Enclosure Award of 1846, over one and a half acres of land there were granted to William Driffield. In 1856, the land was put up for sale and set out with roads and building plots. It was in effect a southerly extension to the resort and became known as New Brighton, after the fashionable resort in Sussex. In 1878, the Local Board of Health noted that roads on the land had not yet been named, nor were the houses numbered. Residents' addresses were given simply as 'New Brighton'. The Board agreed to consult residents on what names to use for the roads.

The main street of the development was initially called College Street. Two colleges owned land in the vicinity so was it called after Trinity College or Sidney Sussex College? Whichever it was, the name did not last long and was changed a month later to Brighton Street – as it is still called. The two other streets were given names which they still have. South Street because it is south of Humber Street. And finally, Hope Street. This latter name was presumably chosen by its hopeful residents; who also appeared to trust in providence according to the stone on the archway between nos 2-4 which is inscribed 'Providence Terrace 1872'. *See also* the separate entry for Grannies Passage, which runs from the end of Hope Street. It is noteworthy that seven other places in this country also acquired the popular name New Brighton, plus more in the United States of America, and probably elsewhere.

NEW CLEETHORPES

New Cleethorpes was that part of the town lying either side of Grimsby Road between Park Street and Manchester Street. Sidney Sussex College developed this area for housing from the mid-1880s. The designation New Cleethorpes was being used by 1895, when it was said that: 'There are two different races of people in Cleethorpes. Those in the south portion wished to go to bed at ten, whilst those in the north wanted to stay up till eleven.' It acquired some notoriety in 1899 when it was reported that: 'In New Cleethorpes there are lamentable signs of feminine drunkenness'. The area was divided from the resort by open fields and most of its workers were employed in Grimsby. Thus it was noted in 1907 in the local press that: 'Cleethorpes Township is divided into two portions known as the "Old" and the "New" Cleethorpes. New Cleethorpes is quite modern and forms a residential suburb of Grimsby'.

NEW ROAD

This was completed in 1879 as a new road off Sea View Street – hence its name.

NEWSTEAD ROAD

The names on a large part of the Cleethorpes Council's extensive Beacon Hill estate are on the theme of castles, stately homes, abbeys, etc. Newstead Abbey in Nottinghamshire was the ancestral seat of the poet Lord Byron.

NICHOLSON STREET *see* WHITE'S ROAD

NORFOLK LANE

Mention of Norfolk Cottages was made in the 1871 census and a directory of 1872. The name Norfolk Lane was given on the 1887 Ordnance Survey map.

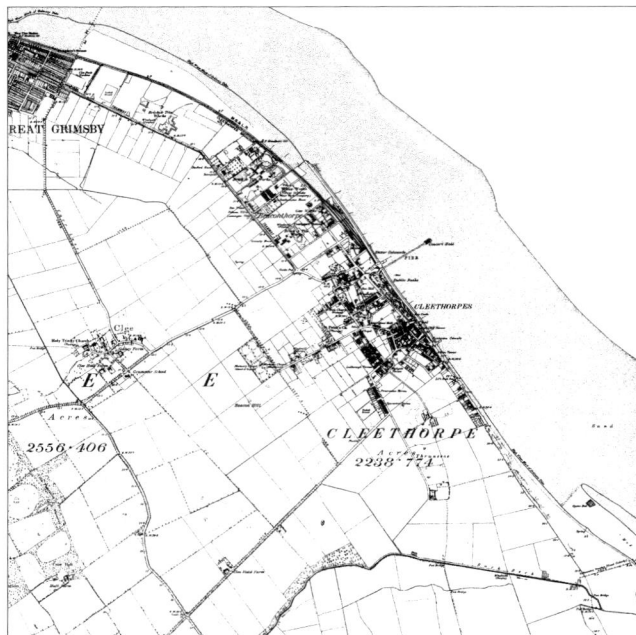

65 *Cleethorpes, 1888-90. By this date, building had taken place largely as infilling or along existing streets and roads. The main areas of development had been the central seafront district and the Beaconthorpe area to the north. In the extreme north-west there is the first slight indication of Grimsby housing spreading over the boundary at Park Street. Compare this with the 1946 map.*

66 *Cleethorpes, 1946. Compared with the 1888-90 map, infilling in the older parts of the town had intensified plus the major expansion of building on to new streets. The town was now largely built up from the Grimsby boundary to Cromwell Road in Thrunscoe.*

NORMANDY ROAD *see* MONTGOMERY ROAD

NORTH STREET

This street still has a terrace of houses bearing the name 'Saunby's Buildings'. There is a story that the terrace was owned by 'Ma Saunby' who also had a farm. Reputedly, she was so well known for watering the milk from her farm that locals christened her houses 'Blue Milk Terrace' (blue milk being skimmed or watered milk). The name passed into common use until, in 1891, the local authority officially settled on the name of North Street – it being a northerly continuation of South Street. The practice of watering milk also occurred in Grimsby; in 1918, a Grimsby grocer was fined £6 for selling milk which contained over 25 per cent added water.

NORTH SEA LANE

This lane was once wholly in the parish of Humberston. However, since 1921 it has marked the southern boundary of Cleethorpes; but only the houses on the northern side of the lane are in Cleethorpes. It acquired its name because for centuries there were two lanes in Humberston leading to the sea marshes. The southern one became South Sea Lane. The northern became North Sea Lane, which has had a chequered history. It was like a cart track in the early 1920s with camp sites located at its seaward end. They were used by summer campers, many in First World War ex-army bell tents which were pre-erected by the camp owners. Further inland along the lane, land was being used for temporary dwellings such as huts, caravans and old converted vehicles. Some had no sanitary conveniences and hedges and ditches were being fouled.

When the lane started to be built up with housing, this mix of temporary accommodation and permanent dwellings caused problems. In 1933, one resident complained to the local authority of the putrid smells from a nearby caravan and of the 'filthy and vile language' used by the occupants. Such problems began to be sorted out by the local authority in the 1930s and we now have an agreeable residential road – with a touch of seaside anticipation at its seaward end.

OOLE

The hamlet of Oole was one of the three thorpes of Clee, the others being Itterby and Thrunscoe, which all eventually grew together to form modern Cleethorpes. The thorpes were outlying hamlets of the parent village of Clee (now known as Old Clee). Oole was recorded in the Lindsey Survey of 1115-18 as 'Hol' meaning a hollow. It has had many different spellings since then, including Hoole (1443) and Howle (1538) but in the Enclosure Award of 1846 it was given its present spelling. It was centred on the present market place and has also been known locally as Fore Thorpe, Low Thorpe and Near Cleethorpes. There were 14 families living there in 1563.

OOLE ROAD

Two roads have borne this name. The earlier one got the name because it led from Itterby to Oole and was known as Oole Road in 1846. Using modern street names, the road ran along Cambridge Street to the junction of St Peter's Avenue and Mill Road. It then turned right and ran along St Peter's Avenue until it arrived in Oole at Short Street. The name was resurrected in the late Victorian/Edwardian period, to be used for the single-sided street which runs between Cambridge Street and Yarra Road. This newer Oole Road once hosted a small sub-fire station at the back of the present Town Hall.

ORMSBY CLOSE

Most of the roads in this corner of the town have names with a local or Lincolnshire theme. Take your pick of the villages North and South Ormsby. The nearest is North Ormsby with its grassed hillside field, which can be seen from the road, and has fascinating 'humps and bumps' and lines marking the layout of a deserted medieval village.

OSBORNE STREET

The street did not exist at the time of the Enclosure Award in 1846 but its line follows the boundary between plots of land held respectively by the Vicar of Clee and Richard Thorold. John Osborne owned land elsewhere in the town in 1846 and land was bought and sold after the Enclosure Award of that year. So it is conceivable that Osborne acquired more land and had Osborne Street constructed.

OSLEAR CRESCENT

Named after the late Alderman Jack Oslear, one-time Chairman of the Cleethorpes Council Housing Committee.

OXFORD STREET

Another puzzling name. The street was known as Oxford Street in 1908 when its building was getting under way; and one of the Edwardian terraces has a stone bearing the inscription 'Asquith Terrace 1909'. At that time Herbert Asquith, later the Earl of Oxford, was prime minister. But he was not created Earl of Oxford until 1925, too late for the street to be named after his ennoblement. However, Asquith was a student at Oxford University, so perhaps that's the connection. Alternatively, did the developers just choose a 'superior' name to give their new road a prestigious quality to match the existing Cambridge Street? The street was subsequently extended towards Cromwell Road and more building took place in the 1920s and 1930s.

67 A snowy Oxford Street, looking down the hill.

OYSTER COURT

This pleasant, recently-built enclave resurrects memories of one of Cleethorpes' major products – the Cleethorpes oysters. They were raised in extensive oyster beds or pits on the foreshore until pollution in the Humber caused production to be ended in 1904.

⊰⊱⊷⊶

PARK STREET

This street, which marks part of the Cleethorpes boundary with Grimsby, was named after the Clee Park Pleasure Ground. This was a commercial leisure ground which hosted a variety of sporting and other recreational activities. It was closed in 1889 when its lease from Sidney Sussex College came to an end and the College developed the site for housing. The land now includes such streets as Taylor Street and Montague Street. The Clee Park was commemorated in the name of the *Clee Park Hotel* which was erected on the corner of Park Street and Grimsby Road in 1890 and demolished in 1990.

Park Street was also the boundary between the areas covered by two police forces – the Grimsby borough constabulary and, in Cleethorpes, the Lincolnshire county constabulary. This led to the tongue-in-cheek explanation that Park Street was so-named because officers of the two police forces would move helpless drunks and other inert bodies to the other side of the street and so 'park' them in the other force's area.

68 *Corner of Park Street (to the right) and Grimsby Road (to the left). Shop decorated for the opening of the Kingsway in 1906.*

69   Clee Park Hotel, *built 1890 and seen here in 1985. It used to mark the boundary between Cleethorpes and Grimsby until its demolition in 1990.*

PARK VIEW

Originally part of Tiverton Street, it was re-named Park View in 1908 because it overlooked Sidney Park, which had been opened in 1904.

PARKER STREET *see* SHERBURN STREET

PARRIS PLACE

Dr Francis Sawyer Parris was Master of Sidney Sussex College, 1746-60. He also bequeathed £400 to the College for the purchase of clerical livings. In 1747, he was sent a valentine by a junior Fellow which he did not appreciate. Accordingly, 'the Master made complaint against Mr Wood for writing and sending him on or about the 14 of Feb. a false, scandalous and abusive letter'. Mr Wood apologised, 'promising for the future to carry myself towards the Master and all others with more decency and respect'. 1933 saw the naming of the cul-de-sac and the commencement of house building. The College acknowledged that residents might be puzzled by the spelling of the street name, which appears to be a mis-spelling of the French capital city.

PEARSON ROAD

Dr Edward Pearson was Master of Sidney Sussex College, 1808-11. He was a former Fellow of the College who held a country vicarage and initially declined when asked to stand for election as Master in 1807. The following year he agreed and was elected. He died after two and a half years in the post. He was highly esteemed as a preacher and was a model parish priest. He attended each week-day at his church 'for the purpose of reading portions of the Liturgy and expounding passages of Scripture and rigorously presiding over the moral and religious education of the children'. In 1811 he had an apoplectic seizure (a stroke) while walking in the garden of his parsonage, and died a few days afterwards.

In Cleethorpes we should also acknowledge a tenuous connection with J.L. Pearson, the architect of Truro Cathedral. He designed the College's student accommodation building, Cloister Court, which was completed in 1891 and which was largely paid for out of the leasehold ground rents received by the College from its Cleethorpes estate.

PELHAM ROAD

Pelham is the family name of the Earls of Yarborough but the Pelhams did not own the land where this road was built. However, the Earl of Yarborough was Lord of the Manor of Cleethorpes-with-Scartho so the name may have been used because of this connection. The road accommodated the town's water works. And subsequently housed a tram depot and an electricity generating station, ready for the inauguration of the electric tram service in 1901. The distinctive water tower in the road was erected in 1908.

PENDRETH PLACE

Charles Pendreth was a Fellow and Tutor at Sidney Sussex College at the time of the Civil War, when the university was mainly Royalist. During the war, Charles I requested a loan from the university. Accordingly, in 1643, Pendreth was one of the Fellows who ordered that £100 (a good sum of money in those days) should be taken from the College Treasury to be sent to the king. In 1657, Pendreth bequeathed to the College 'effects to the value of £83 8s. 5½d.' This went towards the purchase of Evanis Hall at Polstead in Suffolk. The Cleethorpes Council agreed to the cul-de-sac's name in 1933.

PENSHURST ROAD

Built on Sidney Sussex College land, the road was named after Penshurst Place which is the Sidney family's stately home in Kent. In the reign of Henry VIII, Sir William Sidney commanded the right wing of the English army at Flodden. In 1513, he became the king's chamberlain and steward of the royal household. He also attended Henry VIII at the spectacular Field of the Cloth of Gold assembly near Calais in 1520. Penshurst Place was granted to him by Edward VI in 1552. One of Sir William's daughters was Lady Frances, the founder of Sidney Sussex College.

PERSEVERANCE TERRACE *see* THRUNSCOE ROAD

PHELPS STREET

Dr Robert Phelps was Master of Sidney Sussex College, 1843-90. He died in office aged 84. He is a very important figure in the history of Cleethorpes because he was the College's main driving force behind the development of northern Cleethorpes, which became known as New Cleethorpes (*see also* the entry under that name). During the Victorian

70   *Pier Gardens, probably early 1950s.*

71   *The pier in Edwardian times, with the 1905 pavilion to the right.*

72   *The Pier Approach between the two world wars. It was all altered in the late 1930s and the central area is now part of a widened Sea Road.*

reform of Oxford and Cambridge universities, he initially supported reform but later became a leader of the die-hards who opposed more radical proposals. When the Prince Consort was installed as Chancellor of Cambridge University in 1847, Dr Phelps went to Buckingham Palace for the ceremony. Unfortunately, Phelps 'who had been out of sorts for some days, was indiscreet enough to venture on a glass of punch after the turtle soup, which so disagreed with him that he had to be got out of the room'.

PHILIP AVENUE *and* PHILIP GROVE

Named after one of the children of Cartledge the builder, along with nearby Ann Grove and Brian Avenue.

PIER

Opened on Bank Holiday Monday, 4 August 1873, the 1,200ft-long pier attracted 2,859 people who paid sixpence a head for admission. Afterwards the admission charge was one penny. On 29 June 1903, the concert hall at the end of the pier was burned down. Its replacement, the present Pier Pavilion, was opened on 10 June 1905. The pier was 'breached' during the Second World War as an anti-invasion precaution. In 1949, the isolated seaward end was demolished, thus reducing the pier to its present length of about three hundred feet.

PIER GARDENS *see* PROMENADES

PINFOLD

A pinfold is an enclosure where stray livestock can be kept until reclaimed by their owner. Agricultural communities had them, including Cleethorpes. Reference has already been made to an old pinfold in the entry for Cambridge Street. A more recent pinfold was sited just about where the drive of 40, Taylor's Avenue lies. The pinfold's 46 square yards belonged to the Cleethorpes Council. The Council refused to sell it to an interested party in February 1923 on the grounds that it was 'still necessary for the Urban District'. However, in October of the same year it agreed to exchange the pinfold for land in Sherburn Street.

So a new pinfold was created in Sherburn Street. But some of its neighbours were not happy in 1927 and complained to the Council that horses in the pinfold had reached over the five-feet-high wall and damaged trellis work and roses. Accordingly, it was decided to increase the height of the walls to six feet. But as that area became more built-up and agricultural land was turned over to housing, the pinfold was given another use. And so it may now be seen in Sherburn Street. It is the small gated and walled enclosure harbouring an electricity supply sub-station about half-way between Parker Street and Lindsey Road.

POPLAR ROAD

This road was initially known as Poplar Street and in 1887 it ran alongside Conyard's brick pit and yard (*see* the entry for Conyard Road). In November 1904, a fire station was built in the road to house the town's new horse-drawn fire engine. The brick pit was closed before the

73 *The old mortuary in Poplar Road, built in 1904 and seen here in 2009.*

74 *Late-Victorian or Edwardian terrace house in Poplar Road, seen here in 2009.*

First World War. It was later filled in and purchased by the Cleethorpes Council for use as a Council yard and depot. Land in the street continued to be used for public purposes. At various times it contained a mortuary, a sewage pumping station, an ARP Decontamination Station, and a new fire station which was opened in 1940. In 1930, the Cleethorpes Council approved plans by builders Wilkinson & Houghton to build 28 houses as a cul-de-sac which was named Poplar Grove in 1931.

PRINCE'S ROAD

Prince Albert, consort to Queen Victoria, died in 1861, two years before Cleethorpes got its first railway station, which was built at the end of this road. So was it named in his memory? Alternative candidates were his son, the Prince of Wales, who visited Grimsby in 1879 or his grandson, Prince Albert Victor, who opened the new promenade improvements in 1885. The road certainly had its present name by 1887 but major building did not get under way there until the 1930s. In 1930, builders Wilkinson & Houghton got approval to build 14 villas. In 1931, plans were approved to build a County Police Station there. In 1932 it was recommended that trees should be planted along the road and in 1937 plans were approved for the Sisters of St Joseph private nursing home; which has been replaced recently by a high-rise block of apartments.

PRINCESS SQUARE *see* SEGMERE STREET

PROMENADES

The resort has three promenades: the North Promenade, the Central Promenade and the King's Parade. The first two are dealt with here; the King's Parade is covered in the entry for the Kingsway and King's Parade. The North and Central Promenades were opened on 2 July 1885 as part of the seafront improvements carried out by the Manchester, Sheffield and Lincolnshire Railway Company. The North Promenade has always been regarded as the 'trippers end'. This was probably due to it being adjacent to the railway station, with those arriving by train having immediate access to the beach, amusements and public houses. Its large-scale 'Wonderland' amusement venue opened with open-air amusements in 1911. In 1921, the venue was bought by entrepreneur George Wilkie who re-arranged the site and created an indoor amusement complex. In 1987, it was put up for sale and the building is now used as a Sunday Market.

75 *North Promenade in Victorian days.*

76 *A crowded North Promenade in 1957.*

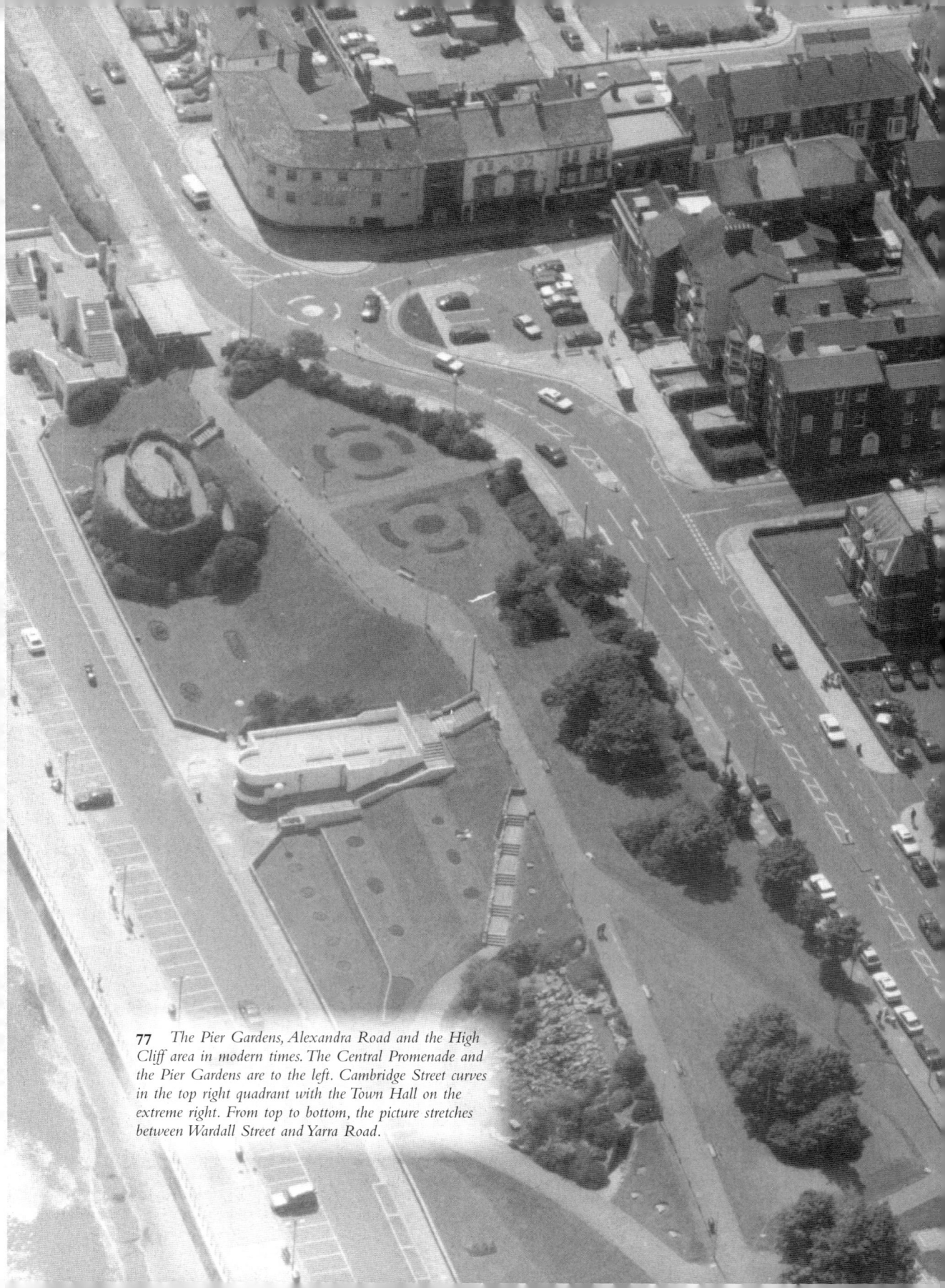

77 *The Pier Gardens, Alexandra Road and the High Cliff area in modern times. The Central Promenade and the Pier Gardens are to the left. Cambridge Street curves in the top right quadrant with the Town Hall on the extreme right. From top to bottom, the picture stretches between Wardall Street and Yarra Road.*

78 *Central Promenade, 1950s.*

The Central Promenade has always been the less hectic promenade, backed as it is by the Pier Gardens. In their early days, the attractions of the promenade and the Pier Gardens included salt-water swimming baths, an aquarium, a camera obscura and cafés. The southern end is still dominated by Ross Castle. This is a 'folly' which serves as an observation point and was constructed in 1885 as part of the railway company's promenade construction works. It was named after Edward Ross (1827-92) who held the important post of secretary of the railway company during 1850-92. Accordingly, he was heavily involved in the management and financing of the company's improvement works at Cleethorpes. He was said to be 'kindly, shrewd and diplomatic. Handsome in appearance and impeccable in his dress, nothing seemed to ruffle him … He also enjoyed immense popularity amongst officers and staff alike'. He died in office.

PYTCHLEY PLACE

Roads in this part of the resort are named after fox hunts in various parts of the country. This one is named after the Pytchley Hunt in Northamptonshire.

✦⟲◯⟳✦

QUEEN MARY AVENUE

It was in 1924 that Sidney Sussex College set out plans to build a 50-feet wide arterial road from Wellington Street in Grimsby to Lestrange Street. The Cleethorpes Council opposed the project on the grounds that it would bring undesirable extra traffic on to Grimsby Road. Opposition to the scheme had declined by 1935 when work on the road began at the Grimsby end and the Council's Highways Committee recommended that the new road should be named Queen Mary's [*sic*] Avenue, after the consort of George V. There were problems later in the decade when the College could not reach agreement with the Grimsby Corporation over the completion of the portion where the avenue connects with Lestrange Street. If the road had been built here to its full width it would have had to go over the resort's boundary on to land which was in Grimsby. Hence the narrowing of the road here. There was extensive house building in the avenue during 1935-9.

QUEEN'S PARADE

This name has been used twice in the resort. In the 1850s part of what is now Alexandra Road was known as Queen's Parade. The name fell out of use and in 1908 it was revived for the new road which had been set out leading from the newly-constructed Kingsway. The nearby promenade had been designated King's Parade, so the name Queen's Parade fitted nicely into this area of royal personas. Approval was given for the building of semi-detached villas there as early as 1909 and plans were approved for over 40 houses during the 1920s and the early 1930s.

QUORN MEWS

Roads in this part of the resort are named after fox hunts in various parts of the country. This one is named after the Quorn Hunt in Leicestershire.

⟶⟶◉⟵⟵

RACE GROUND

Horse races were being held on the marsh land bordering the Humber in the early 1600s. By the 1820s, the 'Cleethorpes Races' were being held in the holiday season on what was called the Horse Course Marsh. A rough race track was laid out on the marshes to the north of the present Grimsby Road. This was certainly no Derby and the main race was usually for 'horses of all ages not thorough bred'. The prize for the winner was usually a saddle. Donkey and pony races were also run, their prizes being whips and bridles.

RAVEN LEYS

This name was given in the 1846 Enclosure Award. It appears to have applied to land on or near the cliff in the vicinity of the junction of Humber Street and High Cliff Road. The reason for the use of the name is unknown but one translation would be the 'land where there are ravens'. It could also be relevant that from this viewpoint, it may have been possible to make out Ravenserodd. This was a port, rivalling Grimsby, that was developed on an island near the mouth of the Humber in the 1230s, but which was destroyed by floods in 1367. *See also* the entry for Fisherman's Road.

RAVENDALE ROAD

One of the cluster of roads south of Chichester Road with names on a Lincolnshire theme. East and West Ravendale lie only a few miles from Cleethorpes on the edge of the Lincolnshire Wolds. The Parkinson family who lived at Ravendale Hall in previous centuries had extensive land holdings in Scunthorpe, where several streets are named after family members.

RECREATION GROUND

In the 1846 Enclosure Award, two and a quarter acres of cliff-top land in the 'Sea Field' were allotted to the 'Churchwardens and Overseers of the said parish of Clee for the use of the Inhabitants of the said Parish and Neighbourhood, as a place of Exercise and Recreation'. This land became known as the Recreation Ground. It ran along the seaward edge of the central cliff and provided a public walk with sea views. The ground was incorporated into the Pier Gardens which were laid out in 1885 during the construction of the Central Promenade. For more information *see* the entry under Promenades.

RESTON COURT

One of the cluster of roads south of Chichester Road with names on a Lincolnshire theme. North and South Reston lie south of Louth on the road to Alford.

REYNOLDS STREET

Three Sidney Sussex College personages are commemorated in this street name. Taking them chronologically, Sir John Reynolds was a student at the College who became a committed Parliamentarian during the Civil War and held commands in Ireland and Flanders. He perished at sea while returning to England. Then we have Richard Reynolds, who was admitted to the College in 1689 at the age of fifteen. He married a daughter of the Bishop of Peterborough and became Dean of Peterborough. In 1721, he became Bishop of Bangor and in 1723 was promoted to Bishop of Lincoln; he died in office in 1744. Finally, we come to John William Reynolds who lectured in history at the College in the period before the First World War. He was highly regarded and was credited with giving a lasting impetus to historical studies at the College. He was killed in the 1914-18 war.

   The street was constructed during 1928-9. In 1933, a resident wrote to the Cleethorpes Council suggesting that trees be planted in the street and that it be renamed 'Way, Avenue, Grove or Road'. The matter was referred to Sidney Sussex College which refused to change the name on the grounds that all the leases had been granted and renaming could lead to confusion in the future. Also, whereas the College was willing to plant trees in new streets it was not prepared to plant them in existing streets. In 1934, the Council agreed to plant trees in the street at a cost of £12.

RICHMOND ROAD

Named after Richmond Castle in Yorkshire, in this small cluster of streets named after castles and stately homes.

RIVERSIDE DRIVE

An understandable name, seeing that the Humber is not too far away.

ROBINSON ROW

This was a group of houses which were once 'on the south east' of Cambridge Street. It included Polly Fewster's cottage, which was one of the places where the Primitive Methodists worshipped in the early 19th century.

ROBSON ROAD

This surname has both an ancient and modern relevance to Sidney Sussex College. In 1648 the College was drawing a useful annual income of about £25 from 'Robson's tenements', which were houses owned by the College in Cambridge. Centuries later, in 1901, H.C. Robson held the important post of College Bursar. As such he had a vital part to play in the administration of the Cleethorpes estate, whose leasehold ground rents were becoming an increasingly significant part of the College's income.

In 1932 the portion of the road from Grimsby Road to Brereton Avenue was adopted by the Cleethorpes Council, followed by the extension to Campden Crescent in 1934. In the meantime, in 1933, the Council received a petition from residents asking for the name to be changed to 'Road'; it was originally called Robson Street. The matter was referred to Sidney Sussex College, which agreed to the request.

ROSS CASTLE see PROMENADES

79 Ross Castle, c.1890.

ROWSTON STREET

The area of land upon which this street lies was put up for auction in 1871 in 28 lots. By 1887, the street extended from the seafront as far as Barkhouse Lane only, with little building evident. At that stage it was named Sussex Street but Sidney Sussex College then gave that name to their street off Park Street. Accordingly, in 1891, the College's local agent and solicitor, W.H. Daubney, asked the Cleethorpes Council to alter the name of the earlier street. So it was renamed Rowston Street, after a long-standing Cleethorpes family. By 1906, the street was built-up with housing to about its present length. The street has an interesting selection of name stones, such as Clarence Terrace, Ivy Cottages, Fells Terrace, Buller Terrace 1901 and Branksome Terrace.

RUFFORD ROAD

The names on a large part of the Cleethorpes Council's extensive Beacon Hill estate are on the theme of castles, stately homes, abbeys, etc. Rufford Abbey is not far from Mansfield and was anciently a Cistercian monastery.

RUSSELL COURT *see* BEDFORD ROAD

RYMER PLACE

Thomas Rymer (1641-1713) was a student at Sidney Sussex College. He had a reputation as a critic, scholar, author and historian. He was appointed Historiographer Royal to William III in 1692 at an annual salary of £200. His chief title to fame is his 20-volume *Foedera*. In this major work, he brought together treaties, alliances, and other documents relating to the Crown's dealings with other kingdoms. It 'remains a collection of great value to historians and antiquaries'. Between 1693-8 he spent £1,253 of his own money on the project; a very large sum of money at that time.

* * *

ST HELIER'S ROAD *see* TENNYSON ROAD

ST HUGH'S AVENUE

St Hugh of Lincoln (c.1140-1200) was Bishop of Lincoln from 1192 and started rebuilding the cathedral which had been largely destroyed by fire and earthquake. St Hugh's Avenue is the spinal road of the town's first council housing estate. The two side roads are Saxon Crescent and Westfield Grove. Both of these roads continue the historical theme. The former alluding to Saxon times and the latter to the parish's West Field which lay beyond the present Middlethorpe Road in 1740. The estate was built in the early 1920s in response to the grievous housing shortage after the First World War. The Cleethorpes Council paid Sidney Sussex College £3,250 for the land and built 136 houses there. But by 1922 over 930 applications had been received for the houses, and more were being received. In 1928, the Lindsey County Council's Child Welfare Centre was opened on the estate; the site now contains the premises of an extensive new medical centre.

ST MARY'S CLOSE

Named after the nearby Old Clee parish church, whose full dedication is to the Holy Trinity and St Mary.

ST PETER'S AVENUE

First known as Oole Road, its name began to change after the consecration of St Peter's Church in 1866. A few years later, a row of four new houses was called St Peter's Terrace, then the road became known as Church Road, then St Peter's Road and finally, in 1908, St Peter's Avenue. It has since been extended twice. Initially it ran only from Short Street to Mill Road but in the early 1890s the road was cut through from Short Street to join up with High Street. The other end of the road was extended later to Highgate. Finally, in the 1960s, the local authority purchased and demolished several houses on Highgate, making a gap which enabled the road to link up with Oxford Street. Thus creating the present through-road from High Street to Cromwell Road. The main shopping area of the avenue really began to evolve in the 1920s with the accelerating change of houses into shops, with their front gardens being taken into the pavement.

80 *Looking down St Peter's Avenue from the High Street, c.1900. The houses on the left are now all shops and the spire of the now-demolished Trinity Wesleyan Methodist Church may be seen in the distance.*

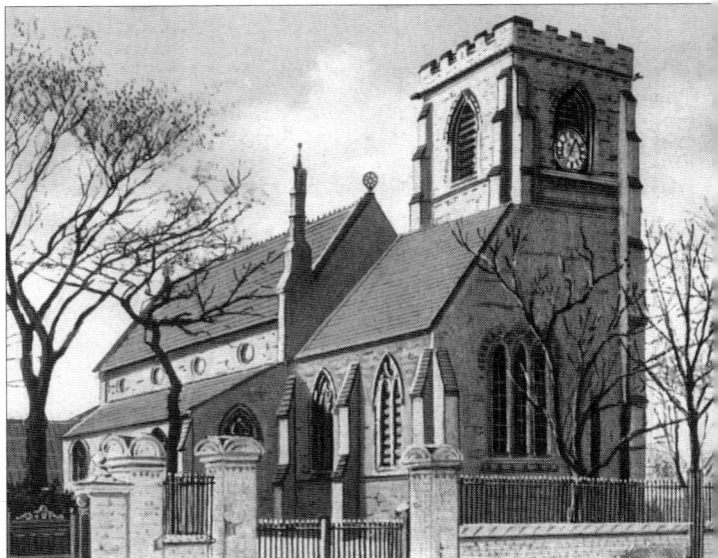

81 *(Left) The Trinity Wesleyan Methodist Church in St Peter's Avenue. It was built in 1885 and its site now contains the shoppers' car park opposite St. Peter's Church.*

82 *(Above) St Peter's Church consecrated in 1866 and seen here in late-Victorian or Edwardian days.*

83 *A tree-lined St Peter's Avenue on a sunny day, possibly in the 1950s. The photographer was standing opposite St Peter's Church and looking towards the High Street.*

84 *The building of Sandringham Road in the 1950s.*

SANDRINGHAM ROAD

The names on a large part of the Cleethorpes Council's extensive Beacon Hill estate are on the theme of castles, stately homes, abbeys, etc. This one is named after the royal residence Sandringham House in Norfolk.

SAUNBY GROVE

In 1931, plans from Mrs B.M. West were approved for 24 houses to be built off Poplar Road. A year later the name Saunby Grove was approved and in 1937 the road was adopted by the Cleethorpes Council. Did this street name have anything to do with the family of Thomas Saunby, Cleethorpes smack owner, who was involved in local building in the 1870s?

SAXON CRESCENT *see* ST HUGH'S AVENUE

SCARBOROUGH STREET *see* CAMBRIDGE STREET

SCHOOL WALK

Its location, backing on to school playing fields, gave this road its name.

SCRIVELSBY COURT

One of the cluster of roads south of Chichester Road with names on a Lincolnshire theme. A couple of miles south of Horncastle, lies Scrivelsby Court, the home of the Dymokes, Lords of the Manor and Grand Champions of England. Up to the time of George IV, the Champion (one of the Dymokes) would ride on horseback into Westminster Hall at the Coronation Banquet and challenge the right of the Sovereign against all comers. Since the Banquet was discontinued the Champion carries one of the Standards in the Coronation ceremony.

SEA ROAD

This major access road to the seafront was already named Sea Road in 1846 when it was described as a 'Public Carriage Road of the breadth of Forty feet' commencing at the junction of what is now Grant Street and High Street before turning down the present Sea Road to the seafront. In 1885 the road was improved and widened as part of the railway company's promenade scheme. In 1938-9, the Cleethorpes Council widened it further, with central traffic islands, and replaced an old Refreshment Room with the present *Submarine* public house. At the time of writing, plans are afoot to redevelop this part of the seafront. *See also* the entry for the Folly Hole, which was the road's earlier name.

SEA BANK ROAD

This ran from Brighton Street southwards along the seafront. Houses were built there including Albion Terrace, which lay between Brighton Street and Rowston Street. In 1871, 13 houses were listed in the terrace, of which 11 were lodging houses and eight of the householders were fishermen.

85   *Sea Road, seen here c.1950s, after the pier was shortened in 1949.*

86 *Sea Bank Road c.1900. Replaced by the Kingsway in 1906. The first road on the left is Bradford Avenue.*

Further along the road (now nos 71-2 Kingsway) was the Grafton College girls boarding school. It was later used as a machine gun school headquarters during the First World War.

When the lifeboat service was inaugurated in 1868 the lifeboat was housed towards the southern end of the road. Its arrival was soon followed by the opening of the *Lifeboat Inn*. Because of difficulties in launching the lifeboat from the foreshore it was moved to Grimsby in 1882 but the inn retained its name. The inn was on the site now occupied by the 'The Waterfront' block of high-rise apartments.

The road was replaced by the Kingsway which was opened in 1906. Before then the houses which lined Sea Bank Road faced on to a muddy track which ran along the top of boulder clay cliffs that were being rapidly eroded. It was recalled in the early 1900s that in the 1870s it had been a broad roadway which had since been eroded by the 'ravages of the tide'. Some houses were in danger of falling into the sea when the work on constructing the Kingsway began in 1903.

87 *The* Lifeboat Hotel *on Sea Bank Road (now the Kingsway) in Victorian days. The site now holds 'The Waterfront' block of high-rise flats on the corner of Queen's Parade.*

SEA FIELD *see* RECREATION GROUND

SEA LANE *see* SEA VIEW STREET

SEA VIEW STREET

An obvious name for this street which overlooks the sea. However, it was referred to as Sea Lane in 1850, but by 1871 had adopted its present name. Up to the late 1880s it ran only between the seafront and Cambridge Street, that is half its present length. Then Sidney Sussex College extended the roadway as far as St Peter's Avenue

88 Sea View Street and the Queen's Hotel, *seen from Cambridge Street and looking towards the sea, c.1900.*

89 Sea View Street looking towards the sea in 1987.

in 1891. The extension was fully built-up to its present length by 1906. Current 'regulars' at the street's *Fisherman's Arms* will shudder to learn that in 1920 the Cleethorpes Council recommended the Licensing Justices to refuse its licence: 'there being three other licensed premises within 250ft of the *Fisherman's Arms*' – but it's still there. The distinctive bank building on the corner of Alexandra Road was approved to be built in 1925.

SEACROFT ROAD

There is a Seacroft a mile south of Skegness. Was it named after this, or was the name used because it has a pleasant seaside-sound about it? The reason why the roadway does not continue through to Cromwell Road may be found in a 1924 petition from 25 residents in Seacroft Road. They asked that the road should be carried through to Cromwell Road as a footpath only. They feared that otherwise their road would be used as a loop road by charabancs and wagonettes proceeding along the Kingsway to the Bathing Pool. Sidney Sussex College also did not want such traffic down their new Cromwell Road and erected a fence to prevent access from Seacroft Road. The Cleethorpes Council finally agreed that Seacroft Road should be continued as a footpath 10 feet wide provided that an open space was retained either side 15 feet deep planted with shrubs, etc. *See also* the entry for Lindum Road.

SEAFORD ROAD

Frank Robinson, the last chairman of the old-established firm of Grimsby trawler owners, Sir Thomas Robinson & Son (Grimsby) Ltd, and who died in Humberston in 2002, had been educated at Seaford School in Sussex.

SEGMERE STREET

This street was known as Princess Square in 1890, as indicated by the stone inscribed 'Princess Square' high on the wall of the corner house. It was probably named after Princess Alexandra, the popular Princess of Wales, who visited Grimsby in 1879. We assume that Alexandra Road was also named after her. The present name of Segmere Street was being used by 1895 and was named after the Segmere drain which once ran along the line of the street and flowed into the Humber. The name Segmere was recorded in 1687 and has been interpreted as meaning 'the pool where sedge [grass] grows'. The drain marked the boundary between Itterby and Thrunscoe.

SHERBURN STREET

There is a collection of streets off Taylors Avenue whose names are a quandary. They are on a piece of land which is bounded on three sides by Taylors Avenue, Trinity Road and Thrunscoe Road. Sherburn Street is the spinal road running the length of the land and the side roads are Woodsley Avenue, Wendover Rise, Lindsey Road, Parker Street and Hey Street. In 1846, the land was divided between Trinity College, Cambridge, and W.N. White; but by the 20th century, the ownership had changed. In 1908, we have mention of the Heys [sic] Estate planning a road layout on the land. By 1925, the Lindsey Lands Investment Co. Ltd was a landowner there, and mention was also made of estate-owners Messrs. Parker of Selby, solicitors.

Although we know nothing of these three landowners, they do give us three definite street names, i.e. Hey Street, Lindsey Road and Parker Street. They also probably give us another. This is because the solicitors at Selby were located only 10 miles from Sherburn-in-Elmet, so a likely connection there would give us Sherburn Street.

With regard to Wendover Rise, Viscount Wendover was one of the titles held by the Carrington family of nearby Humberston. The street runs off Taylors Avenue, which at one time was known as Humberston Road – giving us a possible, if flimsy, connection. Finally, we have Woodsley Avenue. In 1924, Messrs H. & L. Wood appeared on the scene, negotiating with the Cleethorpes Council on the layout of land. Their status is unclear, whether acting as solicitors, agents or owners. But it is possible that the name Woodsley Avenue is an attractive extension of their surname. Some building took place in these streets before the First World War but most of the development occurred in the 1920s and 1930s.

SHORT STREET

The brief length of this street is justification for its name. It is an old street which was there well before 1846.

SIDNEY PARK

The land for the park was given by Sidney Sussex College, hence its name. It was opened in 1904 by Charles Smith, the Master of the College. The original design of the park was by T.H. Mawson (1861-1934) of Windermere who went on to become the leading landscape architect of his day. He worked for many notable clients, including Queen Alexandra and Andrew Carnegie. He designed many public parks and was involved in major town-planning schemes in Canada and Europe. A further four acres were added to the park in 1925, when the Unemployment Grants Commission gave a grant towards the cost of levelling, fencing and drainage work. The additional land was open for public use in April 1926.

90 *Sidney Park, opened in New Cleethorpes in 1904 and seen here in its early days.*

91 *Yachting at Sidney Park, pre-1914.*

SIDNEY STREET

This was named after Sidney Sussex College and was one of the first streets built on land which became New Cleethorpes. Building there started in 1888, when application was made to build six houses in the new street.

SIGNHILLS AVENUE

This name was 'borrowed' from a local geographical feature. 'Sign Hills' was the name given to the area of sand dunes which were later transformed into the Thrunscoe Recreation Ground, with the Bathing Pool, Boating Lake, etc. In 1901, a local newspaper commented on the use of the name. 'Sign' was a local spelling of 'syne' which is marram grass. Accordingly, 'Sign Hills' means sand dunes with marram grass growing on them – this is the tough grass which helps to bind the dunes together. The name also figures in other parts of Lincolnshire; Skegness has a Syne Avenue and Sine Hills were noted at North Somercotes more than a century ago. A small area of the old dunes has been preserved as a nature reserve near the Humberston end of the Boating Lake.

To avoid any misunderstanding, it should be emphasised that Signhills Avenue was not itself built on sand dunes. Indeed, before building took place there it was noticed that the ground was 'hummocky', looking like the site of old-time buildings. As the land is in Thrunscoe, it was possibly part of the site of the Thrunscoe deserted medieval village. *See also* the entries for Lindum Road and Thrunscoe.

SIMONS PLACE

Ralph Simons was the architect of Sidney Sussex College's original buildings of 1595/6. He previously had a reputation in Cambridge as the builder of Emmanuel College and parts of Trinity College. The day-book for the building of Sidney Sussex College contains such references as a workman 'fauling from the scaffold' – before the days of hard hats. In 1934, the Cleethorpes Council approved the street name and also approved plans for the building of 20 houses by Wilkinson & Houghton. The College had originally proposed calling it Hey Place but the Council thought this would cause confusion with the existing Hey Street.

SOLOMON COURT

Councillor Wolfe Solomon was Mayor of Cleethorpes when he welcomed Queen Elizabeth II on her visit to the town in 1958.

SOUTH STREET *see* NEW BRIGHTON

SPURN VIEW TERRACE *see* THRUNSCOE ROAD

STANHOPE PLACE

Stanhope was the family name of Philip, 1st Earl of Chesterfield. He made a donation of £100 to the Sidney Sussex College library. His third son, Ferdinando Stanhope,

92 *Sidney Park, opened in 1904 and seen here in 1963, showing Brereton Avenue running vertically towards the right and Queen Mary Avenue on the extreme left.*

became a Fellow Commoner of the College in January 1635-6. During the Civil War he became a colonel in the king's army, and was killed in action at West Bridgford in Nottinghamshire.

STATION ROAD

The railway came to Cleethorpes in 1863 and in 1878 the Cleethorpes Local Board decided that the road leading from the *Victoria Hotel* (currently *O'Neill's Irish Pub*) to the railway station was to be called Station Road. *See also* the entry for Grant Street.

STRUBBY CLOSE

This was on Sidney Sussex College land but a connection with the College is not readily apparent. The only tentative connection is through the Mountain family who purchased the College's Cleethorpes leasehold estate in the 1960s. John Mountain and W.R. Riggall farmed at Saleby which is only a couple of miles south of Strubby. *See* the entry on Harrington Street for the College's connection with Saleby.

SUGGITT'S LANE

This lane (along with the modern Suggitt's Court and Suggitt's Orchard) were named after the Suggitt family who lived in the lane and had houses built there. There are several confusing stories about the family. Apparently, they were of French extraction and Solomon de Suggitt, was a French senator who was exiled after the French Revolution, whereupon he dropped the 'de' prefix to his name. It is reported that in the 19th century, a member of the family was found guilty of smuggling and divested of his property and land. He was also sentenced to life imprisonment but pardoned after about two and a half years and sent to serve in the Crimean War. He then received a free pardon and was appointed a customs officer at Grimsby.

Then we have Robert Suggitt who was living in the lane in 1880 as a smack owner and whelk merchant. He was born at Easington in East Yorkshire and was about six years old when the family moved to Cleethorpes. He died in 1901 aged 89. Robert's grandfather apparently lived at Kilnsea in a house once raided by men of the infamous privateer Paul Jones. Another Suggitt, William, was born in 1836. He was a well-known smack owner but sold his vessels after they began to be displaced by steam trawlers. He died aged 93 in about 1929.

SUSSEX RECREATION GROUND

The name is taken from Sidney Sussex College which gave 14 acres for a recreation ground in 1919. The land was donated on condition that it was maintained as a public open space and a children's playground. Work was carried out in stages during the 1920s and 1930s, with government financial support on condition that unemployed men were employed. By 1931 the ground contained six hard tennis courts and 10 grass courts, several bowling greens and children's play equipment. The first of the two pavilions was built in 1931 and approval was given in 1934 for the two-storey pavilion to be built.

93 *Major event taking place on the Sussex Recreation Ground, with lots of children and a brass band. In the background looms the bulky 1931 gasholder (demolished 1967).*

94 *Sussex Recreation Ground pavilion, 2009.*

SUSSEX STREET

This was named after Sidney Sussex College and, along with Sidney Street, was one of the first streets built on its New Cleethorpes land. Building there started in 1888, when application was made to build six houses in the new street. There already existed another Sussex Street in the resort's New Brighton area but its name was changed to Rowston Street after a request from the College in 1891.

SWABY DRIVE

Most of the roads in this corner of the town have names with a local or Lincolnshire theme. The hamlet of Swaby is off the A16 road about seven miles south of Louth.

TAYLOR STREET

Samuel Taylor of Dudley became a student at Sidney Sussex College in 1688. In 1723 he bequeathed the College his property in Dudley to promote the study of mathematics. In the following century, mining rights on this property greatly increased the College's income. This in turn facilitated the introduction of a wider range of mathematical and scientific studies at what had hitherto been essentially a theological college.

TAYLOR'S AVENUE

This road has had a confusing succession of names. In 1846, it formed part of Thrunscoe Road. At one time it was known as Clee Field Lane, after Clee Field Farm which lay along the road. It was listed in a local 1923 directory as Humberstone [*sic*] Road but in 1925 it was being referred to as Taylor's Lane. In that year the name was officially changed by the Cleethorpes Council to its present name, Taylor's Avenue. This was despite a local petition asking for it to be called Humberston Road. The petition was rejected because that name had already been used for another road.

In 1928, a Cleethorpes Council committee recommended that permission be given to Mr E. Jackson to establish a Dripping Factory on the south side of the avenue for 12 months. It would be in a brick building which was a good distance from any other building. There was certainly no development on the farther stretch of the avenue at

95 *A rural scene along Taylor's Avenue, looking towards Cleethorpes in the 1950s.*

this time but two events in the same year showed that change was in the offing. Firstly, the tenancy of a cricket pitch on the avenue was terminated because the land had been acquired for building. Secondly, a sub-post office was opened at 8 Taylor's Avenue to cater for the increasing population of the area.

TENNYSON ROAD *and* ST HELIER'S ROAD

These two roads are on land which was part of the Tennyson estate in Cleethorpes and Grimsby. In 1835, the estate was inherited by Frederick Tennyson (1807-98) who was the elder brother of Alfred, Lord Tennyson, the Victorian Poet Laureate. Frederick lived in St Helier, Jersey, during 1859-96, hence the name St Helier's Road. He was also a poet and became increasingly eccentric – once rushing round his house shouting: 'Where are my trousers, where are my trousers? I have forty pairs and can only find thirty-five'.

In 1906-7, leases were granted for 116 terrace houses to be built in St Helier's Road, producing a total annual ground rent of £170. The Beaconthorpe Methodist Church was built on the corner of Tennyson Road in 1914. In 1928, 62 leasehold semi-detached houses were to be built in Tennyson Road producing a total annual ground rent of £142. Then, in 1935, the entire Tennyson leasehold estate in Grimsby and Cleethorpes was put up for auction. The estate was purchased for £115,000 by the trustees of the late Grimsby brewer, W.T. Hewitt.

TERRACES

In past times, many house addresses were given as the name of the terrace in which they were located, rather than the name of the road in which they were situated. Some terraces are featured in entries in this list but for a fuller list *see* Appendix A.

THORESBY PLACE

The names on a large part of the Cleethorpes Council's extensive Beacon Hill estate are on the theme of castles, stately homes, abbeys, etc. Thoresby Park is near Ollerton in Nottinghamshire.

THORGANBY ROAD

One of the cluster of roads south of Chichester Road with names on a Lincolnshire theme. Thorganby is a placid place in the Lincolnshire Wolds whose peace was rudely disturbed during the Civil War when a troop of Parliamentarian soldiers attacked the Hall, shot the steward and had the owner imprisoned at Lincoln.

THORNTON CRESCENT

The names on a large part of the Cleethorpes Council's extensive Beacon Hill estate are on the theme of castles, stately homes, abbeys, etc. Thornton Abbey with its imposing gatehouse built in the 1300s is not too far away and is well worth a visit.

THORPES *see* CLEETHORPES

THRUNSCOE

The hamlet of Thrunscoe was one of the three thorpes of Clee, the others being Oole and Itterby, which all eventually grew together to form modern Cleethorpes. The thorpes were outlying hamlets of the parent village of Clee (now known as Old Clee). The name of this thorpe comes from a Scandinavian compound word meaning 'thorn-bush wood'. Thrunscoe was listed in Domesday Book of 1086, with the spelling Ternescou. Other spellings have been Thirnescho (1212), Thurnsco (1535) and Thrunskoe (1594).

Thrunscoe is that part of the resort which lies between Segmere Street and the Buck Beck. In 1563 sixteen families were living there. Most of them probably lived in an area which stretched along the east side (the sea side) of the present Hardy's Road. By the 1840s only three families were living in Thrunscoe. Accounts of 'hummocky' ground, possibly marking the foundations of past dwellings, have led to Thrunscoe being looked upon as the site of a deserted medieval village. However, house building in the 19th and 20th centuries has covered the land which may have yielded evidence of the old settlement. *See also* the entry for Cleethorpes.

THRUNSCOE LANE *and* THRUNSCOE OCCUPATION ROAD

These were earlier names for Hardy's Road. *See* Hardy's Road for further information.

THRUNSCOE RECREATION GROUND

Named so because it is located in the Thrunscoe part of the resort. The concept of having a recreational facility there was seen by the Cleethorpes Council as a likely outcome of opening up the southern part of the resort by the building of the Kingsway and the King's Parade promenade. Accordingly, in 1901, the Council had the foresight to buy 33 acres of sand dunes from Sidney Sussex College for £500. The College sold the land under the agreement that it was to be used as a public recreation ground and should remain so forever. Development of the land started after the First World War and the open-air Bathing Pool and the Boating Lake were completed in the 1920s. It now contains the Leisure Centre (which was opened in 1983 and replaced the Bathing Pool), the Boating Lake, the children's sand pit, the paddling pool, the Discovery Centre and the Cleethorpes Coast Light Railway.

96 *The Bathing Pool had other uses apart from bathing. Here is a Beauty Parade at the pool c.1936.*

97 *The Boating Lake and the Discovery Centre, 2009*

THRUNSCOE ROAD

This road is not to be confused with the one-time Thrunscoe Occupation Road, which is now Hardy's Road. The present Thrunscoe Road runs between Highgate and Taylor's Avenue, and named because it led from Itterby to Thrunscoe. Development began in the 1870s with the construction of the two rows of terraced houses which lie either side of the end of Sherburn Street. These rows were called Perseverance Terrace and Spurn View Terrace. The former still bears an original stone on which the inscription 'Perseverance Terrace 1871' is practically eroded away. The terraces got their names from the building companies which had them built, i.e. the Cleethorpes Spurn View Building Company and the Great Grimsby Perseverance Building Company.

It should be noted that the present Thrunscoe Road was once part of a much longer Thrunscoe Road which was described in 1846 as a 'Public Carriage Road thirty feet wide' which began in Highgate, proceeded down the present Thrunscoe Road and then along what is now Taylor's Avenue to the present Hewitt's Circus.

98 *Detached villa in Thrunscoe Road, 2009.*

99 *Thrunscoe, in 2002. The Leisure Centre is at the bottom of the picture with Cromwell Road running diagonally from it up to Hardy's Road. Most of the housing to the right of Cromwell Road was built between 1906 and the 1930s. With some notable exceptions, the housing to the left of the road is largely of 1950s vintage or later.*

TIVERTON STREET

Named after the home town in Devon of Peter Blundell, a great benefactor of Sidney Sussex College. For further information *see* the entry on Blundell Avenue.

TRINITY ROAD

Named after Trinity College, Cambridge, which owned a small amount of land in the town. In 1846, this consisted of 17 acres, of which 15 acres were at what is now the southern end of Trinity Road. In 1925, the Cleethorpes Council considered a new road off Taylor's Avenue 'which will ultimately connect with Beacon Avenue; for the present such street to be called Trinity Road' – and the name has remained.

TWINING PLACE

Thomas Twining was the 'natural' son of the tea merchant Daniel Twining, of the well-known tea brand. Initially sent to work in his father's tea business, he was unhappy there, having a 'passion for books, provided they were not books of business'. He was admitted to Sidney Sussex College as a student in 1755 and turned out to be an excellent classical scholar and a fine musician. He became a Fellow of the College in 1760. Prior to going to the College he had been tutored in Latin and Greek, along with his tutor's daughter Elizabeth; he married her in 1764 and settled down as a clergyman in a country parish. He died in 1804. The plans for this cul-de-sac, and its name, were approved by the Cleethorpes Council in 1932. In the same year, plans were approved for 14 houses to be built there by F.C. Taylor and F.A. Would.

UPPER THORPE *see* ITTERBY

100 *Victoria Terrace with tennis courts, all now demolished. The terrace ran down to the North Promenade from the* Victoria Hotel *(currently* O'Neill's Irish Pub*).*

❖⇒◉⇐❖

VICTORIA TERRACE

This striking Victorian terrace of large houses was built at right-angles to the North Promenade. It was sited on what is now a car park between the *Victoria Hotel* (currently known as *O'Neill's Irish Pub*) and the Station Approach. It is said that it was built to screen off the smoky railway station from the centre of the resort. It had become neglected and run down by the 1970s. Conservationists argued that it should be repaired, improved and retained but it was demolished during 1978-9.

❖⇒◉⇐❖

WALDORF ROAD *see* BEDFORD ROAD

WALTHAM GROVE

The village of Waltham is close enough for a trip to see its fine windmill and other attractions.

WARD STREET

Dr Samuel Ward was Master of Sidney Sussex College, 1610-43. He was reputed to be one of the most influential men in the Cambridge of his day. As a true Puritan he admitted his own faults, one of which was 'my immoderate eating of walnuts and cheese after supper, whereof I did distemper my body'. He was also a staunch Royalist and was imprisoned twice in Cambridge by Parliamentarians. The second time was for several weeks, which undermined his health; he died soon after his release.

WARDALL STREET

This street was in the centre of the ancient thorpe of Itterby; and its name is entangled in confusion. A John Wardall was listed locally in 1761. Then, in 1856, Mary Wardle [*sic*] appeared as a grocer. Methodist historian Dr Frank Baker stated that the street was named after a Methodist grocer's shop – having

101 *Wardall Street, the buildings in the foreground are said to have been fishermen's cottages. Seen here in the 1950s. The cottages have since been refurbished and modernised.*

102   *Wardall Street looking towards Cambridge Street. Seen here c.1955, before demolition.*

previously been known as James Street and, later, Middle Street. In 1901, it was noted that opening out from the street were several 'squares set round with tiny cottages'. The houses were closely packed and were in the area most affected by the severe outbreak of cholera in 1854, which killed 60 people in the resort within three weeks. *See also* the entries for Amos Square, Charles Square and James Buildings.

Under 1930s slum clearance legislation, eight houses in the Wardall Street area were demolished by 1938. The Cleethorpes Council then decided that the Clearance Area should henceforth be kept clear of buildings. All the houses in the close-packed area have since been demolished and the land is now the public car park in Wardall Street.

103   *James Square in the 1950s. The site is now part of the Wardall Street car park.*

WARNEFORD ROAD

This street began to be developed after the First World War and may have been named after a wartime hero who was awarded the Victoria Cross upon becoming the first pilot to shoot down a German Zeppelin. He was Flight Sub-Lieutenant R.A.J. Warneford, RN, who destroyed the airship by dropping six bombs on it over Belgium on 7 June, 1915. The street had already been named by 1920 when land was being sold there for housing. It was fully built up by 1931.

WARWICK ROAD *see* DUDLEY PLACE

WEEKES ROAD

The Rev. George Arthur Weekes was Master of Sidney Sussex College, 1918-45. Consequently, he was in charge of the College during the 1920s and 1930s when it was still heavily engaged in turning its Cleethorpes land over to leasehold housing. He also guided the College through a difficult period of its own history, bedevilled as it was by the effects of two world wars and the nationwide problems of the inter-war period. His devotion to the College was centred on its chapel. He retired at the age of 75 and left the College £1,000, which was used as the nucleus of a fund to keep the College chapel furnished as he would have wished.

WEELSBY ROAD

This road runs from Nuns Corner to Love Lane Corner and in doing so cuts across Weelsby. In general terms the Weelsby Woods public park marks the approximate centre of Weelsby, which was absorbed into Grimsby in 1889. Hitherto, Weelsby had been in the parish of Clee.

It should also be noted that in 1846, Weelsby Road was the name given to the road which began at Balls Land (at the southern end of Ladysmith Road) and ran up to and beyond Love Lane Corner (along what is now Clee Road) to the bottom of Isaac's Hill. It then continued up to the top of the hill as far as the *Cross Keys* public house, which was where the car park is now on the High Street.

WELBECK ROAD

Streets in this small corner assembly are named after prestigious places which have castles, cathedrals or stately homes. Welbeck Abbey is near Worksop in Nottinghamshire.

WENDOVER RISE *see* SHERBURN STREET

WESLEY CRESCENT

One of the cluster of roads south of Chichester Road with names on a Lincolnshire theme. This one honours a renowned county personage, namely John Wesley, the founder of Methodism. He visited Cleethorpes in 1781 and met with an enthusiastic reception.

WESTBURY ROAD *and* **WESTBURY PARK** *see* **BEDFORD ROAD**

WESTFIELD GROVE *see* **ST HUGH'S AVENUE**

WHITEHALL ROAD

Named after the Whitehall Farm of about 184 acres which lay on the northern side of North Sea Lane.

WHITE'S LANE *see* **CORONATION ROAD**

WHITE'S ROAD

The names of this road and Nicholson Street derive from the landowner, William Nicholson White, who once owned the land where these roads were being set out for building in 1906.

WILLIAM STREET

The land here was awarded to Eliza Sleight in the 1846 Enclosure Award. In 1906, only the stretch from Mill Road to just beyond Edward Street had been built. 1932-3 saw the street being extended to Highgate by Taylor & Coulbeck.

WILSON STREET

This street was named by 1920. Was it named after Professor C.T.R. Wilson (1869-1959) who was a student at Sidney Sussex College in the 1880s and became a Fellow of the College in 1900? He was one of the 'remarkable band of physicists working in the Cavendish Laboratory, Cambridge, at the turn of the century, at the time of the discovery of the electron, x-rays and radioactivity'. He was awarded the Nobel Prize for Physics in 1927.

WINDSOR ROAD

Streets in this small corner assembly are named after prestigious places, such as Windsor, which have castles, cathedrals or stately homes.

WINN COURT

Joshua ('Josh') Winn was mayor of Cleethorpes in 1963, 1979 and 1991. He retired from the Council in 1995; having served, with a break of only three years, since 1956. He died in 2003.

WOLLASTON ROAD

The Wollaston family had a strong allegiance to Sidney Sussex College. At least 10 members of the family studied there. William Wollaston became a student in 1674 and later sent his four sons to the College. Then one of these sons sent his four sons there. During 1807-8, William's great-grandson, Professor Francis Wollaston, was Master of the College. He combined scientific teaching with a career in the church. Later in life he devoted himself to the service of the church and became an Archdeacon. The first applications to build in the road were made in 1905.

WOODSLEY AVENUE *see* SHERBURN STREET

WORKING MEN'S

This 'nickname' was applied to Bancroft Street, George Street and Charles Street and arose because the houses there were built by the Cleethorpes Working Men's Building Society. The streets were completed and dedicated in 1882. Bancroft Street ran between Charles Street and Highgate. It is now part of St Peter's Avenue. George Bancroft was listed locally as a brickmaker in 1856: this gives us the names of two of the streets. The other one, Charles Street, is probably named after another member of the family. George Street and Charles Street were later extended; for example, in 1933 plans were approved for 46 houses to be built in George Street.

⊷⟝◉⟞⊶

YARRA ROAD

The road was first shown as an 'intended new road' on the sale plan of 'very eligible building land' in 1844. Yarra Yarra is the name of a river in Australia which runs into the sea at Melbourne. A fishing smack called the *Yarra Yarra* and owned by Henry Knott was up for sale locally in 1875. It had been built at Burton Stather in 1867. The imposing

104 *Yarra House, on the corner of Alexandra Road and Yarra Road. The site now contains the public library's garden on Alexandra Road.*

Yarra House once stood on what is now the library garden on the corner of Alexandra Road. The house was built by Joseph Chapman. The road still contains the resort's old post office with a 1907 datestone. Motorists may like to note that the charge for parking in the 'emergency stand' on the road in June 1926 was two shillings and sixpence for charabancs and one shilling and sixpence for cars. *See also* the entry for Itterby Road.

YORK PLACE

Streets in this small corner assembly are named after prestigious places, such as York, which have castles, cathedrals or stately homes.

YOUNG PLACE

John Young entered Sidney Sussex College in 1606 from St Andrew's University. He was Dean of the College in 1612 and later became Dean of Winchester. He was the son of Sir Peter Young, diplomat and tutor to James I, and was said to be the first Scotsman to take a degree in Cambridge. The cul-de-sac is one of several streets off Queen Mary Avenue, the plans of which were approved by the Cleethorpes Council in 1939.

Appendix A
Terraces

Along some of the older streets of the town, stones in the walls of houses are inscribed with the names of terraces or the names of rows of two or more villas, etc. In earlier times, these names frequently served as the sole address of a house, the name of the road (if it had a name) being omitted. The name stones which have survived are part of the local heritage and add interest to the current street scene. They can also be an important tool in historical research. Unfortunately, many have been obliterated or covered over with modern screeding. Consequently, the following list of name stones indicates the location of those which are still to be seen; plus those which are no longer evident but of which a location, approximate or otherwise, is known.

The list does not claim to be complete and any additional information would be welcome. Another list may be consulted on pages 26-9 of *The Place-Names of Lincolnshire, Part Five: The Wapentake of Bradley,* by Kenneth Cameron, which may be consulted in the Grimsby Central Library's Reference Library.

*   *The asterisks indicate those terraces, etc., of which there is no remaining visible evidence.*
*   Albion Terrace, Sea Bank Road.
   Alpha Terrace, on nos 8-10 West Street.
   Asquith Terrace 1909, on nos 26-8 Oxford Street.
   Astley Terrace, on no. 93 Poplar Road.
   Barkfield Terrace, on no. 35 Humber Street.
   Bateman's Avenue 1888, on no. 40 Albert Road.
*   Birmingham Villas, built in 1872 as three 'stately villas', now nos 22-3 Kingsway and the adjacent café on the corner of the Kingsway and Segmere Street.
   Bradford Terrace 1876, on no. 2 Bradford Avenue.
   Branksome Terrace, on no. 26 Rowston Street.
   Buller Terrace 1900, on no. 29 Rowston Street.
   Cambridge Terrace 1874, on nos 13-15 Cambridge Street.
   Claremont Terrace, on no. 1 Isaac's Hill.

Clarence Terrace, on no. 1 Elm Road.

Clarence Terrace, on nos 3–5 Rowston Street.

Clee Crescent, on nos 17–19 Clee Road (partially screeded over).

Dowsing Crescent, on nos 96–8 Mill Road.

Fells Terrace, above archway between nos 21–3 Rowston Street.

Firle Cottages, on no. 43 Humber Street.

Gordon Villas, on nos 40–2 St Peter's Avenue.

* High Cliff Terrace, High Cliff Road.

Itterby Terrace 1875, archway between nos 24–32 Humber Street.

Ivy Cottages, on nos 15–17 Rowston Street.

John's Terrace, archway between nos 83–5 Mill Road.

King George V Coronation Homes 1911, Mill Road.

Ludborough Terrace A.D.1861, on nos 62–4 Highgate.

Lyndhurst Terrace, on nos 120–2 Mill Road.

Lyndon Cottages 1885, De Lacy Lane.

* Mantell's Buildings, on Itterby Road [Alexandra Road] in the region of Yarra Road.

Maygrove Terrace, on no. 4 Highgate and on the Oxford Street return wall.

* Nottingham Terrace, Cambridge Street.

* Oakhurst Terrace, was on no. 1 St Peter's Avenue.

Olive Villas, above archway between nos.57–9 Highgate.

Perseverance Terrace, on nos 40–2 Thrunscoe Road (stone severely weathered).

Poplar Villas, on nos 53–5 Suggitt's Lane.

Pretoria Villas 1902, on nos 39–41 Highgate.

Princess Square, in Segmere Street on return wall of no. 26 Kingsway.

Providence Terrace 1872, on archway between nos 2–4 Hope Street.

* Queens Parade, Alexandra Road.

* Robinson Row, Cambridge Street.

St James Terrace, on no.71 St Peter's Avenue (stone now indistinct).

* St Peter's Terrace, St Peter's Avenue.

Sandhurst Terrace, on nos 13–15 Humber Street.

Saunby's Buildings, on no. 8 North Street.

* Sidney Terrace, High Street, along the corner where it meets Grant Street.

* Sidney Terrace, Grimsby Road, south west side, along the corner where it meets Park Street.

* Spurn View Terrace, Thrunscoe Road.

Victoria Terrace, Suggitt's Lane, on archway between nos 72–4.

* Victoria Terrace, Sea Road. (*See* entry in the main list of streets).

APPENDIX B
Footpaths, 1846

In times past, when most local journeys were made on foot, public footpaths were important. Accordingly, the Enclosure Award of 1846 carefully set down the routes of major local paths, which it specified had to be at least six feet wide. Their routes are described below, using modern street names. But bear in mind that most of the streets or roads which are mentioned below would not be there in 1846 – just the footpaths or tracks.

THE CLEE *and* GRIMSBY FOOTPATH

This path ran downhill from the top of Isaac's Hill, turned left on to Clee Road and carried on to Old Clee, where it continued through the village. Then from the end of Church Lane it ran by the playing fields of the Old Clee Primary School and finished at a point which is now about half way along Ladysmith Road. This latter section is still heavily used.

THE HUMBERSTON FOOTPATH

This path is still in regular use and is also said to have been known as Love Lane. It is the path that runs from the cottages at Love Lane Corner to meet up with the aforementioned Clee and Grimsby Footpath.

THE GRIMSBY MARSH FOOTPATH

This ran from the bottom of Isaac's Hill along Grimsby Road and then continued along Cleethorpe Road in Grimsby as far as the old Grimsby boundary at Humber Street.

THE THRUNSCOE *and* HUMBERSTON FOOTPATH

This path had a complicated route. It ran from the junction of Cambridge Street and Highgate; then along Highgate as far as Thrunscoe Road. Here it turned left and continued along Thrunscoe Road and Hardy's Road until it reached what was Hardy's Farm. There it turned left (through what are now the grounds of the Signhills Schools) until it reached Chichester Road (this point is marked by the gated path which leads out of the school grounds on to Chichester Road). The path then went along where Links Road now runs and terminated at the footbridge over the Buck Beck which gives access to the Country Park. The footpath continued in Humberston, following the line of the present path through the Country Park which comes out at Bedford Road.

THE SCHOOL FOOTPATH

This path began in the Market Place and went along Mill Place as far as Albert Road. It then continued along the Cuttleby and ended at Cambridge Street.

THE HUMBER BANK FOOTPATH

This was described as 'commencing in Cleethorpes at the Fisherman's Road and extending in a Southerly direction along the Humber Bank until it enters the Parish of Humberston near the New Clough'. There have been extensive changes in this part of the resort over the years; including marsh drainage, new sea banks, new roads and the development of the Thrunscoe Recreation Ground. Accordingly, all we can say is that the footpath commenced at the bottom of Highcliff (by Brighton Street) and made its way along the then sea bank as far as the Buck Beck – beyond which lay Humberston.

THE SCARTHO FOOTPATH

This path ran roughly along what is now Vaughan Avenue and then behind Hunsley Crescent to what was described as 'the Hamlet of Weelsby' – now part of the Weelsby Woods public park. In Weelsby it linked up with a longer footpath which would come out in the present Grantham Avenue in Scartho.

THE ITTERBY FOOTPATH

This was a wider path, eight feet wide, which ran from the *Dolphin Hotel* along what is now Alexandra Road and into Sea View Street.

TWO SEA ROAD FOOTPATHS

These two paths were also wider, each being eight feet wide, and ran either side of the Sea Road 'to the shore of the Humber'.

Appendix C
Sidney Sussex College and Street Names

By the time Sidney Sussex College had finished developing its Cleethorpes estate, it had given names to about sixty streets, plus a park and a recreation ground. The College policy was that all the names had to be associated in some way with the College.

Accordingly, its first two new streets, begun in 1888 off Park Street, were named Sidney Street and Sussex Street after the College and its founder, Lady Frances Sidney Sussex. Later, in November 1889, the College's local agent and solicitor, W.H. Daubney asked what names it wanted to use for the continuations of the existing Thorold, Spencer and Stirling Streets (which had all been built by A.W.T. Grant-Thorold on his land in New Clee). The College replied that the continuations of these streets on to College land should be called Harrington, Taylor and Montague Streets respectively – again after College notables.

In the following century, the Cleethorpes Council became more involved in the naming of streets, as may be seen in its convoluted discussions with the College regarding the naming of Brereton Avenue; which are outlined above in the main entry on the avenue.

Sometimes, the College's choice of names was not appreciated by residents. A case in point occurred in 1933 when a letter from its local agent and solicitor, by now H. Mountain, reported that he had received complaints about 'the rather unusual names the College are naming their Streets'. It was the projected name 'Chafy Place' which had brought the matter to a head and Mountain wrote to the College on 6 March 1933 that:

> Unusual names are difficult to remember, and in some respects have to do with the popularity of the houses in the road. This may seem very childish to you but I can assure you it does affect the type of people who are buying these small houses, and I am writing privately to you to see if anything can be done towards having the names more commonplace in the future.

The College replied on 13 March that it had now decided to change the name to Frankland Place and wrote that:

> Your letter has disturbed us a little, for we are wondering if there is any objection to any of the other names for the cul-de-sacs. I suppose Pendreth Place may be said to be difficult and perhaps the two 'r's in Parris is all wrong!

The naming of streets, etc. in Cleethorpes is taken by us quite seriously and the names we use are in some way or other associated with the College, such as Benefactors, past-Masters and distinguished Fellows, etc.

I suppose the authorities do not object to a cul-de-sac being called a Place. If they do, what do you suggest? Is the trouble merely that the names used appear outlandish? Obviously the resident of Cleethorpes looks at these names from a different angle to ourselves.

Mountain replied on 10 May that:

Matters seemed to have settled down now since Chafy Place was re-named … there is certainly no objection to cul-de-sacs being called Place and … this seems to be a more suitable title for a cul-de-sac than Street … We do not regard the matter seriously, but on the other hand if the College can select names which are easy to remember, and which are free from any extraordinary spelling we think it would be advisable. The only thing we can suggest is that before names are finally decided upon, you might submit several to us, and we could find out whether there would be any objection.

A request was made to the College later in the year to change Robson Street and Reynolds Street to 'Road'. The College replied on 17 October that it had no objection to Robson Street being re-named 'Road' and that:

What we must clearly do in the future is when any more naming of streets is required we must send them to you to ask if the Council approve. That may prevent such things as are now occurring.

The request regarding Reynolds Street was turned down. Also turned down, in 1935, was a request to the College for a name change for Fisher Place. More details about these two requests are given above in the main entries for the two streets.

Possibly, the finding of suitable College-related names became more difficult as the decades passed and in May 1958 the College Meeting 'agreed to appoint annually a Names Committee to prepare a list of suggested names for new roads on the College Estate in Cleethorpes'. However, only 10 years later, the College had sold off its Cleethorpes estate and it was perhaps with a sigh of relief that the College authorities realised that the naming of streets was no longer their problem.

(Sources: Sidney Sussex College Muniment Room, Cleethorpes Correspondence Files).

Sources & Bibliography

ARCHIVAL SOURCES

Grimsby Central Library, Local History Collection,
  Town Hall Square, Grimsby, DN31 1HG
  - for books, periodicals, local newspapers, directories, maps, photographs, and other material.
North East Lincolnshire Archives Office,
  Town Hall Square, Grimsby, DN31 1HX
  - for minutes of the Cleethorpes Local Board, the Cleethorpes Urban District Council and the Cleethorpes Borough Council; planning records; the 1846 Clee Enclosure Award and maps; and other papers.
Sidney Sussex College Muniment Room,
  Sidney Sussex College, Cambridge, CB2 3HU
  - for minutes of the Sidney Sussex College Meetings, Cleethorpes leases and correspondence; and other papers.

PRINTED MATERIAL AND FURTHER READING

Ambler, R.W. and Dowling, A., 'The Growth of Cleethorpes and the Prosperity of Sidney, 1616-1968', in Beales, D.E.D. and Nisbet, H.B. (eds), *Sidney Sussex College, Cambridge: historical essays in commemoration of the Quatercentenary* (1996)

Aspinall, P.J., 'Speculative Builders and the Development of Cleethorpes, 1850-1900', in *Lincolnshire History and Archaeology, Vol.II* (1976) pp. 43-52

Baker, F., *The Story of Cleethorpes and the Contribution of Methodism* (1953)

Brookes, R.J., *The Streets of Grimsby and Cleethorpes* (n.d.) (Typescript in Grimsby Central Library Reference Library)

Cameron, K., *The Place-Names of Lincolnshire, Part Five: The Wapentake of Bradley* (1997)

Dobson, E.A., *Guide and Directory to Cleethorpes with an historical account of the place* (1850)

Dobson, E.A., *New Guide and Directory to Cleethorpes with an historical account of the place* (1858)

Dowling, A., *Cleethorpes: the Creation of a Resort* (2005)

Drury, E., *The Old Clee Story* (n.d.)

Edwards, G.M., *Sidney Sussex College* (1899)

Foster, C.W. and Longley, T., *The Lincolnshire Domesday and the Lindsey Survey* (1924)

Hart, M., *Cleethorpes and the Meggies* (1981)

Kay, A., *Aspects of Cleethorpes in the Nineteenth Century and Resort Development, using contemporary documents* (1987)

Mee, A., *Lincolnshire*, revised and edited by F.T. Baker (1970)

Pevsner, N. and Harris, J., *Lincolnshire*, revised by N. Antram (1989)

Room, A., *The Street Names of England* (1992)

Russell, E. and R.C., *Landscape Changes in South Humberside: the enclosures of thirty-seven parishes* (1982)

Scott-Giles, C.W., *Sidney Sussex College* (1975)

Thorold, H. and Yates, J., *Lincolnshire* (1965)

Watson, C.E., *A History of Clee and the Thorpes of Clee* (1901)